Industry 4.0 and Climate Change

At present, both Industry 4.0 and industrial engineering management developments are reshaping the industrial sector worldwide. Industry 4.0 and sustainability are considered as the crucial emerging trends in industrial production systems. The resulting transformations are changing production modes from traditional to digital, intelligent, and decentralized. It is expected that Industry 4.0 will help drive sustainability in industries thanks to the implementation of advanced technology and a move towards social sustainability.

This book reflects on the consequences of the transition to Industry 4.0 for climate change. The book presents a systemic overview of the current negative impacts of digitization on the environment and showcases a new outline of the energy domain and expected changes in environmental pollution levels under Industry 4.0. It also analyzes the ecological consequences of the growth and development of Industry 4.0 and considers Industry 4.0 as an alternative to fighting climate change, in the sense of shifting the global community's attention from environmental protection to consolidation of the digital economy. This book will be of interest to academicians and practitioners in the fields of climate change and development of Industry 4.0, and it will contribute to national economic policies for fighting climate change and corporate strategies of sustainable development under Industry 4.0.

Science, Technology, and Management Series

Series Editor: J. Paulo Davim, Professor, Department of Mechanical Engineering, University of Aveiro, Portugal

This book series focuses on special volumes from conferences, workshops, and symposiums, as well as volumes on topics of current interests in all aspects of science, technology, and management. The series will discuss topics such as mathematics, chemistry, physics, materials science, nanosciences, sustainability science, computational sciences, mechanical engineering, industrial engineering, manufacturing engineering, mechatronics engineering, electrical engineering, systems engineering, biomedical engineering, management sciences, economical science, human resource management, social sciences, and engineering education. The books will present principles, models, techniques, methodologies, and applications of science, technology, and management.

For more information about this series, please visit: www.routledge.com/Science-Technology-and-Management/book-series/CRCSCITECMAN

Industry 4.0 and Climate Change

Edited by

Rajeev Agrawal, J. Paulo Davim,
Maria L.R. Varela, and Monica Sharma

CRC Press
Taylor & Francis Group
Boca Raton London New York Leiden

CRC Press is an imprint of the
Taylor & Francis Group, an **informa** business

A BALKEMA BOOK

Cover image: Shutterstock

First published: 2023
by CRC Press/Balkema
Schipholweg 107C, 2316 XC Leiden, The Netherlands
e-mail: enquiries@taylorandfrancis.com
www.routledge.com – www.taylorandfrancis.com

CRC Press/Balkema is an imprint of the Taylor & Francis Group, an informa business

Library of Congress Cataloging-in-Publication Data
Names: Agrawal, Rajeev (Professor of mechanical engineering), editor. | Davim, J. Paulo, editor. | Leonilde Rocha Varela, Maria, editor. | Sharma, Monica (Professor of management), editor.
Title: Industry 4.0 and climate change / edited by Rajeev Agrawal, J. Paulo Davim, Maria L.R. Varela, Monica Sharma.
Description: Boca Raton : CRC Press, 2023. | Series: Science, technology, and management series | Includes bibliographical references and index.
Subjects: LCSH: Industry 4.0—Environmental aspects. | Green technology. | Sustainable development.
Classification: LCC T59.6 .I35 2023 (print) | LCC T59.6 (ebook) | DDC 658.4/0380285574—dc23/eng/20220808
LC record available at https://lccn.loc.gov/2022017683
LC ebook record available at https://lccn.loc.gov/2022017684

ISBN: 978-1-032-27631-1 (hbk)
ISBN: 978-1-032-27635-9 (pbk)
ISBN: 978-1-003-29357-6 (ebk)

DOI: 10.1201/9781003293576

Typeset in Times New Roman
by codeMantra

Contents

Editor biographies

Dr. Rajeev Agrawal received his ME in 2001 from MNNIT, Allahabad, and PhD in 2012 from BIT, Mesra. He is presently working as Associate Professor in the Department of Mechanical Engineering at MNIT, Jaipur. He has more than 20 years of teaching and research experience. Dr. Rajeev Agrawal has been actively involved in research & development (R&D) activities in the engineering space. He is actively involved in bringing industry orientation to the engineering education system in India working with several industries, statutory bodies and other organizations. Dr. Rajeev Agrawal's currently managing research projects include Sustainability in Manufacturing, Lean Six Sigma, Supply Chain Design and Reconfigurable Manufacturing System (RMS). One of the objectives of his current research is to address conversion quickly for the production of new products by providing customized flexibility and can be improved, upgraded and reconfigured in response to fluctuating demands for automotive companies situated in India. Dr. Agrawal has published more than 70 research articles in leading international journals and international conferences including the *Computers and Industrial Engineering, International Journal of Advanced Manufacturing Technology*, and *Journal of Enterprise and Information Management.* He has authored/co-authored more than 10 book chapters in leading books including *Sustainable Manufacturing and Intelligent Manufacturing.* He is serving as an editorial board member of *International Journal of Business and System Research* and an active reviewer for more than 30 journals from publishers like Springer, Sage and Elsevier. His current research studies include sustainable supply chain, machine learning, optimization algorithms, life cycle assessment, Industry 4.0 and manufacturing systems.

Dr. Monica Sharma is Head and Associate Professor in the Department of Management Studies at the Malaviya National Institute of Technology, Jaipur. She is also the coordinator of the MNIT Innovation and Incubation Centre (MIIC), a DST-GOI-sponsored Technology Business Incubator sanctioned by MHRD, looking after all the activities related to incubation and effective management of the TBI at MNIT, Jaipur. She earned her BE in Mechanical Engineering from Rajasthan University; earned MBA in Production and Operations Management from FMS Udaipur, Rajasthan; and completed PhD in Operations Management from the Birla Institute of Technology and Science, Pilani, Rajasthan. She is a certified quality engineer from IIQM Jaipur and is a certified six sigma green belt from ISI Delhi. She has also undertaken various projects from UGC, DST (GoI, New Delhi). She has

organized more than 30 workshops/stttp/fdp related to entrepreneurship development, sustainable manufacturing, etc. She has supervised more than six PhDs along
with four PhDs ongoing and has written several books as well. She is also a member of IIIE, member of AIMS International, life time member of Indian Institute
of Industrial Engineering and member of Indian Society of Ergonomics. She has
published several journal articles in reputed journals like *International Journal of
Occupational Safety and Ergonomics*, *Journal of Cleaner Production* and *The TQM
Magazine* along with research articles in international conferences. She is also a
reviewer of various reputed journals like *Measuring Business Excellence* and *Lean
Six Sigma* since last five years. Her research interests are industrial engineering,
lean manufacturing, materials management, operations management, operation
research, total quality management, manufacturing excellence, sustainable manufacturing and Six Sigma.

Dr. Maria Leonilde Rocha Varela received her PhD degree in Industrial Engineering
and Management from the University of Minho, Portugal, in 2007. She is Assis-
tant Professor in the Department of Production and Systems at the University of
Minho. Her main research interests are in Manufacturing Management, Production Planning and Control, Optimization, Artificial Intelligence, Meta-heuristics,
Scheduling, Web-based Systems, Services and Technologies, mainly for supporting
Engineering and Industrial Management, Collaborative Networks, Decision-Making Models, Methods and Systems, and Virtual and Distributed Enterprises. She
has published more than 150 refereed scientific papers in international conferences
and in international scientific books and journals, and indexed in the Web of Science and/or in the Scopus data bases. She coordinates R&D projects in the area
of Production and Systems Engineering, namely concerning the development of
web-based platforms and decision support models, methods and systems. She is a
frequent paper reviewer for several journals, such as the *Journal of Computer Integrated Manufacturing*, *Journal of Decision Systems*, *International Journal of Management and Fuzzy Systems*, *International Journal of Sustainability Management and
Information Technologies*, *International Journal of Intelligent Enterprise*, *International Journal of Decision Support Systems Technology*, *Management and Production
Engineering Review*, *Mathematical Problems in Engineering* and *Journal of Robotics*.
Moreover, she has been improving international scientific collaborations, mainly in
terms of projects and special issues proposals and publications with colleagues from
other national and foreign institutions, such as VSB-Technick Univerzita Ostrava;
University of Belgrade; International Islamic University Malaysia; SimTech Simulation Technology; Malek Ashtar University of Technology; University of Macedo-
nia; ED&F Man Commodities West Africa; University of Liverpool; Universidad
de La Frontera; University of Poznan; Toulouse University; University of Liberec;
University of Plymouth; Dalian University of Technology; VIT University; Erciyes
University, Kayseri; University of Calgary; and China University of Petroleum. Her
main scientific activity regarding international conferences is related to the participation as a member in organizing and scientific committees, namely in HIS-Hybrid
and Intelligent Systems; NaBIC – World Congress on Nature and Biologically In-
spired Computing; CaSoN–International Conference on Computational Aspects
of Social Networks; EWG– ICDSST-The International Conference on Decision

Support System Technology; ISDA – The International Conference on Intelligent Systems Design and Applications; WorldCist – World Conference on Information Systems and Technologies; SoCPaR – International Conference on Soft Computing and Pattern Recognition Technologies and Tools; WICT – World Congress on Information and Communication Technologies; IAS – The International Conference on Information Assurance and Security; IBICA – the International Conference on Innovations in Bio-Inspired Computing and Applications; FCTA – International Conference on Fuzzy Computation Theory and Applications; MECAHITECH – International Conference on Innovations, Recent Trends and Challenges in Mechatronics, Mechanical Engineering and New High–Tech Products Development; Manufacturing-International Scientific-Technical Conference; MESIC – Manufacturing Engineering Society International Conference; IJCCI – International Joint Conference on Computational Intelligence; and ViNOrg – International Conference on Virtual and Networked Organizations Emergent Technologies and Tools. Also noteworthy is her participation in international networks, such as Machine Intelligence Research Labs, Scientific Network for Innovation and Research Excellence (MirLabs), Euro Working Group of Decision Support Systems (EWG-DSS), Institute of Electrical and Electronics Engineers (IEEE); System, Man, and Cybernetics Society (IEEE SMC), Industrial Engineering Network (IE Network), and Institute of Industrial and Systems Engineers (IISE).

Prof. J. Paulo Davim received PhD degree in Mechanical Engineering in 1997, MSc degree in Mechanical Engineering (materials and manufacturing processes) in 1991, Mechanical Engineering degree (five years) in 1986 from the University of Porto (FEUP), the Aggregate title (Full Habilitation) from the University of Coimbra in 2005 and the DSc from London Metropolitan University in 2013. He is Eur Ing by FEANI, Brussels, and Senior Chartered Engineer by the Portuguese Institution of Engineers with an MBA and Specialist title in Engineering and Industrial Management. Currently, he is Professor in the Department of Mechanical Engineering of the University of Aveiro, Portugal. He has more than 30 years of teaching and research experience in Manufacturing, Materials and Mechanical Engineering with special emphasis in Machining and Tribology. He also has interest in Management and Industrial Engineering and Higher Education for Sustainability and Engineering Education. He has guided large numbers of postdoc, PhD, and master's students and coordinated and participated in several research projects. He has received several scientific awards. He has worked as an evaluator of projects for international research agencies and an examiner of PhD thesis for many universities. He is Editor-in-Chief of several international journals, Guest Editor of journals, books Editor, book Series Editor and Scientific Advisory for many international journals and conferences. Presently, he is an editorial board member of 25 international journals and acts as a reviewer for more than 80 prestigious Web of Science journals. In addition, he has also published as editor (and co-editor) for more than 100 books and as author (and co-author) for more than 10 books, 80 chapters and 400 articles in journals and conferences (more than 200 articles in journals indexed in Web of Science core collection/h-index 45+/6000+ citations and SCOPUS/h-index 52+/8000+ citations).

List of contributors

R. S. Aakhash
Department of Production Engineering
National Institute of Technology
Tiruchirappalli, India

Bappa Acherjee
Department of Production and Industrial Engineering
Birla Institute of Technology
Mesra, Ranchi, India

Ayaz Afsar
Department of Mechanical Engineering
Government Polytechnic College
Amravati, India

Deepak Agarwal
IET
Dr Ram Manohar Lohia University
Awadh, Uttar Pradesh, India

Mayank Agarwal
IET
Dr Ram Manohar Lohia University
Awadh, Uttar Pradesh, India

Ambeesh Mon S.
Institute of Management in Kerala
University of Kerala
Thiruvananthapuram, India

A. Arora
Mechanical Engineering Department
Delhi Technological University
New Delhi, India

S. Awasthi
Mechanical Engineering Department
Delhi Technological University
New Delhi, India

Pramod Belkhode
Department of Mechanical Engineering
Laxminarayan Institute of Technology
Nagpur, India

Gouraw Beohar
Department of Mechanical Engineering
Shri Ram Institute of Technology
Jabalpur, India

D. Bhargava
Mechanical Engineering Department
Delhi Technological University
New Delhi, India

Tara Charan Bharti
Department of Operations
Tata Steel Downstream Products Limited
Jajpur, India

Pratyush Bhatt
Mechanical Engineering Department
Delhi Technological University
New Delhi, India

Kanchan Borkar
Department of Mechanical Engineering
MES
Nagpur, India

Abhijit Das
Department of Management Studies
Indian Institute of Technology (ISM)
Dhanbad, India

K. Dharun Prashanth
Department of Production Engineering
National Institute of Technology
Tiruchirappalli, India

Immanuel Edinbarough
Industrial and Manufacturing Engineering Department
The University of Texas
Austin, Texas

Aakash Ghosh
Mechanical Engineering Department
Delhi Technological University
New Delhi, India

Sushovan Ghosh
Productivity Services Department
Tata Steel Limited
Jajpur, India

Ravi Prakash Gorthi
Department of Computer Science and Engineering
The LNM Institute of Information Technology (LNMIIT)
Jaipur, India

Navriti Gupta
Mechanical Engineering Department
Delhi Technological University
New Delhi, India

Hareesh Kumar U. R.
Institute of Management in Kerala
University of Kerala
Thiruvananthapuram, India

Jerin Joseph
Industrial Engineering Department
Siddaganga Institute of Technology
Tumkur, Karnataka, India

Akarsha Kadadevaramath
Mechanical department
Intel India Pvt. Ltd
Bangalore, Karnataka, India

Rajeshwar Siddeshwar Kadadevaramath
Industrial Engineering Department
Siddaganga Institute of Technology
Tumkur, Karnataka, India

Mithilesh Kumar
School of Life Science
Jaipur National University
Jaipur, India

Anita Kumari
Department of Production and Industrial Engineering
Birla Institute of Technology
Mesra, Ranchi, India

B. Lathashankar
Industrial Engineering Department
Siddaganga Institute of Technology
Tumkur, Karnataka, India

Chittaranjan Mallick
Department of Mathematics
Parala Maharaja Engineering College
Berhampur, Odisha, India

Y. B. Mathur
Department of Mechanical Engineering
Government Polytechnic College
Bikaner, India

Rajesh P. Mishra
Department of Mechanical Engineering
Birla Institute of Technology and Science
Pilani, India

Sephali Mohanty
Department of Mathematics
C. V. Raman Global University
Bhubaneswar, Odisha, India

Sandeep Mondal
Department of Management Studies
Indian Institute of Technology (ISM)
Dhanbad, India

Soumil Mukherjee
Department of Mechanical-Mechatronics Engineering
The LNM Institute of Information Technology (LNMIIT)
Jaipur, India

K. Muthu Narayanan
Department of Production Engineering
National Institute of Technology
Tiruchirappalli, India

Ramji Nagariya
School of Business and Management
Christ (Deemed to be University)
Delhi-NCR, Ghaziabad, India

Anjaney Pandey
FET
Mahatma Gandhi Chitrakoot Gramodaya Vishwavidyalaya
Chitrakoot, India

Shatrudhan Pandey
Department of Production and Industrial Engineering
Birla Institute of Technology
Mesra, Ranchi, India

P. Parthiban
Department of Production Engineering
National Institute of Technology
Tiruchirappalli, India

Bharat Singh Patel
Department of Operations
Thiagrajar School of Management
Madurai, India

Rama Kant
Mechanical Engineering Department
Delhi Technological University
New Delhi, India

S. Ranjan
Mechanical Engineering Department
Delhi Technological University
New Delhi, India

Sudhansu Sekhar Routary
Department of Mathematics and Humanities
College of Engineering and Technology
Bhubaneswar, Odisha, India

V. Roy
Mechanical Engineering Department
Delhi Technological University
New Delhi, India

Mohan Sangli
Industrial Engineering Department
Siddaganga Institute of Technology
Tumkur, Karnataka, India

Aryan Sharma
Mechanical Engineering Department
Delhi Technological University
New Delhi, India

Deepak Sharma
Department of Mechanical Engineering
Engineering College
Bikaner, India

Priyanshu Sharma
Department of Management
Birla Institute of Technology
Mesra, Jaipur Campus, India

Rahul Sharma
Department of Mechanical Engineering
Poornima College of Engineering
Jaipur, India

Satyendra Kumar Sharma
Department of Management
Birla Institute of Technology and Science
Pilani, India

Rajkumar Sharma
Department of Management
Birla Institute of Technology and Science
Pilani, India

Sumer Sunil Sharma
Productivity Services Department
Tata Steel Limited
Jajpur, India

Vikram Sharma
Department of Mechanical-Mechatronics Engineering
The LNM Institute of Information Technology (LNMIIT)
Jaipur, India

S. Shaw
Mechanical Engineering Department
Delhi Technological University
New Delhi, India

Abhishek Kumar Singh
Department of Production and Industrial Engineering
Birla Institute of Technology
Mesra, Ranchi, India

Anurag Singh
IET
Dr Ram Manohar Lohia University
Awadh, Uttar Pradesh, India

Chanchal Singh
IET
Dr Ram Manohar Lohia University
Awadh, Uttar Pradesh, India

Dharmendra Singh
Department of Mechanical Engineering
Engineering College
Bikaner, India

Trailokyanath Singh
Department of Mathematics
C. V. Raman Global University
Bhubaneswar, Odisha, India

Prachi Swain
Department of Mathematics
C. V. Raman Global University
Bhubaneswar, Odisha, India

Pranav Taneja
Mechanical Engineering Department
Delhi Technological University
New Delhi, India

R. S. Walia
Mechanical Engineering Department
Delhi Technological University
New Delhi, India

Prashant Washimkar
Department of Mechanical Engineering
Government Engineering College
Chandrapur, India

Chapter 1

Optimization of milling machine parameters by using Artificial Neural Network model

D. Bhargava, Ramakant, and Navriti Gupta

CONTENTS

1.1   INTRODUCTION

In recent times, many commercial producers have taken steps to increase the efficiency, accuracy, and durability of their machining products. These steps involve increasing the quality of merchandise, lowering production value, and improving the manufacturing rate. These steps may be influenced by different factors like machining parameters, tool geometry, and the cloth of the painting's piece. The turning process is influenced by input parameters as cutting speed, feed rate, depth of cut, cutting conditions etc. Therefore, the favored finish floor is normally targeted and the ideal strategies are selected to attain the specified first-rate. Generally, those models have a complicated courting among floor roughness and operational parameters, painting materials, and chip breaker sorts. Till now, many researchers have executed experimental investigations approximately the machining operations and evaluated the impact of machining parameters on the outputs of the manner; many researchers have tried to model the machining approaches by using diverse strategies like statistical, shrewd, and analytical techniques [1,2]. Maximum predictive fashions are able to estimate complicated relationships between machining input parameters and corresponding outputs [3,4].

Therefore, those methods notably lessen the required experimental tests for prediction outputs of the technique. Artificial Neural Networks (ANN), fuzzy logic (FL), and regression fashions are a few instances of these strategies [5,6]. Among them, ANN is one of the most widely used methods for presenting a predictive model of the machining method. ANNs were educated to clear up non-linear and complex issues that

DOI: 10.1201/9781003293576-1

aren't exactly modeled mathematically. Many researchers have succeeded to calculate various outputs at detailed enter parameters with the use of ANN. So, this technique can be a useful tool for minimizing the timing and value of the commercial field.

1.2 LITERATURE REVIEW

Turning, like other mechanical techniques, is the process of removing fabric which is an important function in many production strategies within a business. This process has a significant impact on the overall cost of the product. In response, space finishing depends on a variety of factors including food price, material housing, operational complexity, slowing down, density, reduction of time, and the radius of the tool nose. In line with those parameters, a comprehensive literature study was conducted as follows: M. S. Sukuma et al. [7] focused on testing the development and optimization of parameters using the Taguchi and ANN method to the value of Al 6061. Yanis [8] used the response surface methodology (RSM) and ANN tools in predicting the Solid Carbon SS400 Metal Utility using an end-to-end mill. Crude carbide EMC5410, where input parameters reduce speed, feed in line with its coating and axial reduction. Soleymani Yazdi and Khorram [9] see modeling and improvement of the grinding process using RSM and ANN methods in 6061-T6 Aluminum where the parameters include feed charging, spin speed, and cutting depth and cutting parameters are low deviation (roughness value, Ra) and termination (Material Removal Rate, MRR). Ajith Arul Daniel et al. [10] improved multi-cause prediction and performance parameters of the aluminum hybrid steel matrix grinding machine ANN application and Taguchi-gray-related experiments where the expected number of test statistics proposed by the ANN model is more significant than the retrospective version. Mahdavinejad et al. [11] expect more complexity with the help of enhancing the digestive parameters using the neural artificial community and the artificial defense device when using the FB4MB digestive gadget, 40 mm wide and 6 inserts and the R390-1806 12M-PM tool used in Ti -6Al. The 4 V fabric with the insertion parameter reduced the speed, coating of each of them, and the depth of the cut where the results show that in order to have a large finished space within the grinding process, the feed rate should be as low as possible. Natarajan et al. [12] predicted and analyzed non-metallic features of the non-metallic material using ANN during computer and numerically controlled (CNC) rotation when the compound was changed into a pattern using time training (TR), a hundred absorbing scale to measure the normal weight (Ra) values of all templates and determine the price feed becomes the most influential parameter, followed by rotation and cutting depth. Mundadaa and Naralab [13] improved digestive function of ANN use and found that the upper parameter that regulates the intake of food corresponds to the speed assistance, radial rake attitude, and nasal radius. Ajith Arul Daniel et al. [10] improved machine parameters.

First, it used the Taguchi S/N measurement method to investigate the control effect of the entire response parameter of the response; second, it used ANN and Decline to predict the parameters which were really interesting. Input parameters have been large fractions, cutting particle size, depth reduction, feed speed and cut and output predicted MRR, maximum folding (Ra), temperature (T), Feed Force (FF), Radial power component (Fr), and Tangential pressure (toes). Gupta et al. [14] applied ANN-based development at the same time as the complex turn of EN 31. In any of the other research activities, ANN was used to improve machine parameters [15,16].

1.3   METHODOLOGY

1.3.1   Data collection from the experimental setup

Data was collected from the experimental setup. In the experimental setup, we have used a milling machine to perform the experiment. The workpiece material used in this study is hard steel (EN31). Parameters were used in many of the research papers such as cutting speed, depth of cut, feed rate, and the parameter used for output or decision parameters such as surface roughness.

These parameters were selected for analyzing and optimizing data input parameters to get optimum output.

1.3.2   Developing Artificial Neural Network model

The ANN is an information processing model that is based on brain processes in sequence. It contains a large number of interconnected processing elements (called neurons) that work together to solve a specific problem. A major component of ANN neurons is synthesized. Each neuron receives input from several other neurons, multiplies them by assigned weights, and increases and transmits the amount to one or more neurons. Some synthetic neurons may activate the output function of the output before transmitting it to the next variant. Neural processing networks consist of an input layer, which receives data from external sources and one or more hidden layers that process data, and an output layer that provides one or more data points based on network processing (Figure 1.1).

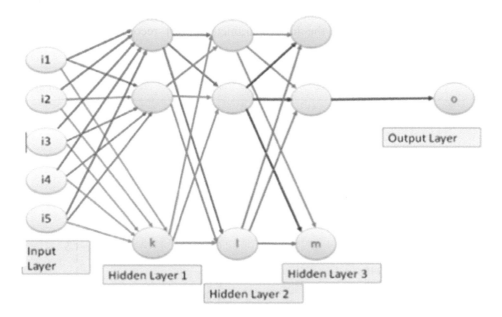

Figure 1.1 ANN-based neural network [14].

In the current activity, a feed-forward back-propagation training algorithm is used to predict high difficulty in the turning process. Training begins with all the weights placed on random numbers. In each data record, the predicted value is compared to the desired (real) value and the weights are adjusted to bring the prediction to the required value.

Many cycles are done with a whole set of training data weights that are continuously adjusted to produce more accurate predictions.

1.3.3 Execution of the experiment

The data set in Table 1.1 was analyzed using ANN.

1.3.4 Prediction using ANN tool

Table 1.2 depicts a comparison between experimental and ANN-predicted results.

Figure 1.2 depicts the ANN plots showing the training, validation, test, and training data. The closeness of R values to 1 shows the high accuracy of ANN-based model.

The MATLAB® form in Figure 1.3 denotes the conditions at which the model is prepared.

The median error is −0.007179847454 which indicates that the improved model has good accuracy in predicting higher difficulty values.

The model used here is very accurate.

1.4 RESULT AND DISCUSSION

Actual roughness values are calculated across all study sets and are equal compared to the expected difficulty values. How to use the selected variable formats calculated in the name of the input parameters? Test records are analyzed by

Table 1.1 ANN based Neural Network

Run	Cutting speed	Feed rate	Depth of cut
1	140	0.05	1.5
2	185	0.04	1.5
3	185	0.04	1.5
4	140	0.061	1.5
5	240	0.061	1
6	240	0.04	1.5
7	240	0.05	1.5
8	140	0.04	1
9	185	0.061	1
10	240	0.061	1.5
11	185	0.05	1.5
12	140	0.04	1.5
13	240	0.04	1
14	240	0.05	1
15	185	0.04	1
16	185	0.061	1.5
17	140	0.05	1
18	140	0.061	1

Table 1.2 Experimental vs. ANN-predicted results

Run	Cutting speed	Feed rate	Depth of cut	Surface roughness (Ra) μm	Predicted	Error
1	140	0.05	1.5	0.59	0.6001982401	-0.0001982401084
2	185	0.04	1.5	0.49	0.6299950013	-0.2199950013
3	185	0.04	1.5	0.59	0.6299950013	5.00E-06
4	140	0.061	1.5	0.79	0.8399736004	2.64E-05
5	240	0.061	1	0.85	0.7358587598	0.08414124017
6	240	0.04	1.5	0.44	0.4547241687	0.005275831313
7	240	0.05	1.5	0.77	0.7991243519	0.0008756480941
8	140	0.04	1	0.74	0.7456387842	0.01436121577
9	185	0.061	1	0.69	0.6499638692	3.61E-05
10	240	0.061	1.5	0.39	0.4144895414	-0.004489541408
11	185	0.05	1.5	0.92	0.9787248987	0.001275101348
12	140	0.04	1.5	0.45	0.4999808545	1.91E-05
13	240	0.04	1	0.49	0.5298561845	0.0001438155294
14	240	0.05	1	0.54	0.5903998999	-0.000399899915
15	185	0.04	1	0.41	0.4415428001	-0.01154280006
16	185	0.061	1.5	0.79	0.7698917502	0.0001082497948
17	140	0.05	1	0.98	0.9001134642	-0.000113464236
18	140	0.061	1	0.42	0.4387660838	0.001233916208

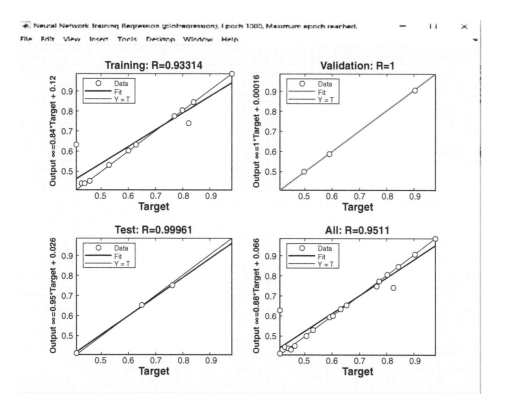

Figure 1.2 Regression plot in ANN.

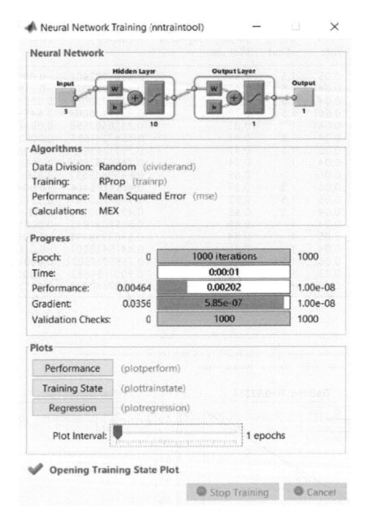

Figure 1.3 Setup for the training environment.

MATLAB_R2019a_v9.6.0.1072779x64 by the ANN strategy. Then the actual grinding figures are compared with the estimates obtained by the use of ANN. Rigid behavior with various parameters is analyzed, and the effect of each parameter over significant value is obtained from the completed test. The difference in the ratio between the actual heavy prices and the estimated revenue figures is obtained at an average of 12.9%.

1.5   CONCLUSION

In this experiment, data of milling machine parameters had some errors, but still, a very good experiment was conducted. Training to testing ratio is 80:20. We successfully did this with the ANN tool with the help of MATLAB.

REFERENCES

1. I. Mukherjee and P. K. Ray, A review optimizations techniques in metal cutting processes, *Computers & Industrial Engineering*, 50 (2006) 15–34.
2. C. X. J. Feng and X. Wang, Development of empirical models for surface roughness prediction in finish turning, *International Journal of Advanced Manufacturing Technology*, 20 (2002) 348–356.
3. A. M. Zain, H. Haron and S. Sharif, Prediction of surface roughness in the end milling machining using artificial neural network, *Journal of Expert Systems with Applications*, 37 (2010) 1755–1768.
4. V. S. Sharma, S. Dhiman, R. Sehgal and S. K. Sharma, Estimation of cutting forces and surface roughness for hard turning using neural networks, *Journal of Intelligent Manufacturing*, 19 (2008) 473–483.
5. A. M. Zain, H. Haron and S. Sharif, Application of GA to optimize cutting conditions for minimizing surface rough-ness in end milling machining process, *Expert Systems with Applications*, 37 (2010) 4650–4659.
6. Y. Jiao, S. Lei, Z. J. Pei and E. S. Lee, Fuzzy adaptive networks in machining process modeling: surface roughness prediction for turning operations, *International Journal of Machine Tools & Manufacture*, 44 (2004) 1643–1651.
7. M. S. Sukumar, P. Venkata Ram, and A. Nagarjuna, Optimization and prediction of parameters in face milling of Al-6061 using Taguchi and ANN approach, *Procedia Engineering* 97 (2014) 365–371.
8. M. Yanis, A. S. Mohruni, S. Sharif, I. Yani, A. Arifin and B. Khona'Ah, Application of RSM and ANN in predicting surface roughness for side milling process under environmentally friendly cutting fluid, series, *Journal of Physics: Conference Series* 1198 (2016) 042016.
9. M. R. Soleymani Yazdi and A. Khorram, Modeling and optimization of milling process by using RSM and ANN methods, *IACSIT International Journal of Engineering and Technology*, 2(5) (2010) 474, ISSN: 1793-8236.
10. S. Ajith Arul Daniel, R. Pugazhenthi, R. Kumar and S. Vijayananth, Multi objective prediction and optimization of control parameters in the milling of aluminium hybrid metal matrix composites using ANN and Taguchi -grey relational analysis, *Defence Technology* 15 (2019) 545–556.
11. R. A. Mahdavinejad, N. Khani and M. M. Seyyed Fakhrabadi, Optimization of milling parameters using artificial neural network and artificial immune system, *Journal of Mechanical Science and Technology* 26(12) (2012) 4097–4104.
12. C. Natarajan, S. Muthu and P. Karuppuswamy, Prediction and analysis of surface roughness characteristics of a non-ferrous material using ANN in CNC turning, *International Journal of Advance Manufacturing Technology* 57 (2011) 1043–1051.
13. V. Mundadaa and S. K. R. Naralab, Optimization of milling operations using artificial neural networks (ANN) and simulated annealing algorithm (SAA), *Proceedings* 5 (2018) 4971–4985.
14. N. Gupta, A. K. Agrawal and R. S. Walia (2019, May), Soft modeling approach in predicting surface roughness, temperature, cutting forces in hard turning process using artificial neural network: an empirical study. In *International Conference on Information, Communication and Computing Technology* (pp. 206–215). Springer, Singapore.
15. N. Gupta and R. S. Walia (2021), Predictive soft modeling of turning parameters using artificial neural network. In R. Agrawal et al. (eds.), *Recent Advances in Smart Manufacturing and Materials* (pp. 189–196). Springer, Singapore.
16. H. D. S. Aiyar, G. Chauhan and N. Gupta (2021), Soft modeling of WEDM process in prediction of surface roughness using artificial neural networks. In R. Agrawal et al. (eds), *Recent Advances in Smart Manufacturing and Materials*(pp. 465–474). Lecture Notes in Mechanical Engineering. Springer, Singapore.

Chapter 2

Facility layout optimization

Continuous improvement

Rajkumar Sharma and Satyendra Kumar Sharma

CONTENTS

2.1 INTRODUCTION

Facility layout can be defined as an understanding of the physical organization of a production system (Meller & Gau, 1996). The layout design of a manufacturing factory does have a significant impact on the success of planning projects (Amar & Abouabdellah, 2016). A well-planned layout makes a more economical and reliable production process (Schenk, Wirth & Müller, 2014). As per the studies, an effective layout can reduce the operating cost by up to 30% (Balakrishnan & Cheng, 2007). Facility layout is a step-by-step process of designing a layout for the flow of material within the plant from raw material to finished goods.

Industrial layouts are a very important factor in the success of a manufacturing plant. There are certain objectives behind the design of an industrial layout like the flow of material, optimization of time utilization, optimum space utilization and safety of equipment, product and people (Meller & Gau, 1996). A lot of work has been done on the optimization of designed layouts also (Nyemba & Mbohwa, 2017; Chandra Mouli, Kulkarni & Deepak, 2021; Wu & Wang, 2017). Optimization techniques like SLP (Systematic Layout Planning), MOM (Multi-Objective Models) and GA (General Algorithm) are being used in the optimization of already designed layouts for improved results for fulfillment of requirements of facility layout (Barnwal & Dharmadhikari, 2016).

DOI: 10.1201/9781003293576-2

In this chapter, we study the designed layout of an electronics equipment manufacturing plant and tried to understand the gaps in the design, the possible impact of gaps and their remedies.

In this case study, the layout is changed using the standard phenomenon of the facility layout of a manufacturing plant. The affected factors are studied and the impact of facility layout is studied. In this chapter, we aim to study and understand the designed layout and tried to find out the improvement opportunities in the layout, planned the act of improvement and did an impact analysis of improved factors.

Novelty of the work: The approach of covering and measuring the variables which play a vital role in the production and are highly affected by facility layout is the novelty of this work. The way of looking into the criticality of layout suggests adding time-to-time review of facility layout in operations and making it part of continuous improvement.

2.2   LITERATURE REVIEW

Objectives of facility layout

1. Optimum space for equipment, space for movement of goods and provide a safe and comfortable work environment to the people as well equipment.
2. Objective base production promotion.
3. Optimum movement of raw material, people and equipment.
4. Flexibility of introducing new lines or technology updating.
5. Increase production capacity of plant." (Kovács & Kot, 2017)

Types of facility layout

1. "Product or line layout
2. Process or functional layout
3. Fixed position layout
4. Cellular type layout
5. Combination layout." (Kovács & Kot, 2017) (Figures 2.1–2.4)

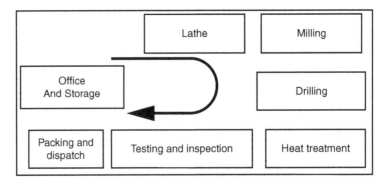

Figure 2.1 Product or line layout.

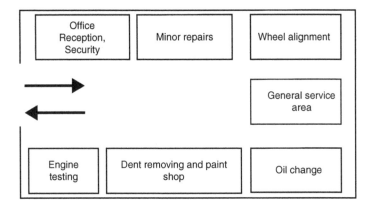

Figure 2.2 Process or functional layout.

Figure 2.3 Fixed position layout.

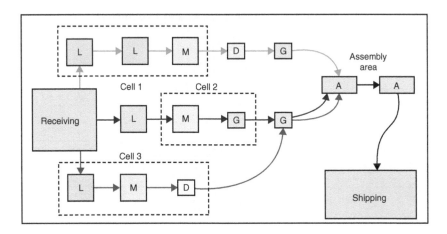

Figure 2.4 Cellular type layout.

Case studies on facility layout improvements

"Case study 1 – conducted in a furniture company in Zimbabwe on facility layout. Changes were done in material flow in the plant. Gains – 43% reduction in movement, improved material flow which ensured the on-time delivery, increased safety and improved quality control." (Nyemba & Mbohwa, 2017).

"Case study 2 – conducted on a small plant by plant layout improvement. Changes were done to the layout and modification of fixtures. Gains – reduction in rework by 10–15% and reduction in material handling by 15%." (Chandra Mouli, Kulkarni & Deepak, 2021).

"Case study 3 – conducted on a chemical industry by improvement in the area-wide layout of plant. Gains – material flow and heat flow had been optimized; environment safety improved." (Wu & Wang, 2017).

There are two identified methods for layout improvement. The first is the routing of material flow for improvement in total material movement distance and time. The other one is a re-layout. Re-layout needs more time and effort and is relatively more expensive (Zhang, Batta & Nagi, 2009).

Lean manufacturing tools and CORELAP algorithm are also used in recent layout optimization work. Non-value-adding activities can be removed just by the study of facility layout by these methods (Tarigan et al., 2019). Graph method and genetic algorithm are used for re-layout design. Displacement of material is easily reduced by this study (Tarigan, Herdita & Tarigan, 2019).

Facility layout optimization is now one of the key requirements because industries are approaching toward EGV in factories which need an optimum path; hence, facility layout optimization is a required key skill in industries (Chen et al., 2019).

2.3   CASE STUDY AND METHODOLOGY

2.3.1   Overview and general layout of the company

The research was carried out in a manufacturing plant of electronic equipment. The plant had a variety of products such as A, B, C, and D named in the layout figure, i.e., Figure 2.5.

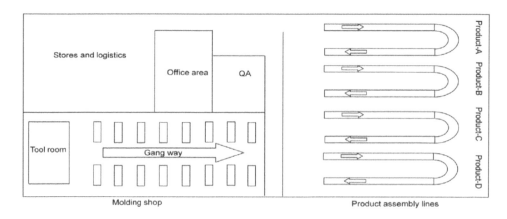

Figure 2.5 Plant layout.

U-shaped integrated assembly lines are there for a variety of products. All assembly lines are grouped on one shop floor named as product assembly shop floor.

There is one molding shop of injection molding machines for in-house plastic parts manufacturing. The molding shop has a tool room in it for handling molds and 34 injection molding machines of different capacities.

The plant has a store for material handling, a QA department and an office area.

Work done on product assembly shop floor –

1. Assembly integration of components
2. Functional testing and packing

Work done on the molding shop floor –

1. Molding of plastic components
2. Assembly of small plastic and metallic parts on some molded plastic parts

In this scenario, the plant layout is designed as per the **product layout**. Product-wise integrated lines are designed.

Grouping of work or process-based layout is missing in the layout facility, which is visible as the second work done in the molding shop.

2.3.2   Gap in the current layout

In the current layout of the plant, the product layout strategy is well-executed but process-based work grouping is missing.

It can be a hybrid plant layout having both process- and product-based layouts. In our study, we planned to study four highly recommended factors in facility layout design with the current scenario and plan to major the impact in the hybrid methodology of plant layout.

Those four factors are

1. Material flow and inventory handling
2. Productivity
3. Quality
4. Safety

2.3.2.1   Factor 1 – material flow and inventory handling

As shown in Figure 2.6, material flow take place from stores to the product assembly shop floor and on the molding shop floor as well. After the molding of plastic components and assembly of small parts on the molded part, that material goes to the product assembly shop floor.

As shown in Figure 2.6, there is a total of eight inventory heads, which are denoted on the arrows in the figure.

Material flow from stores to product assembly shop floor

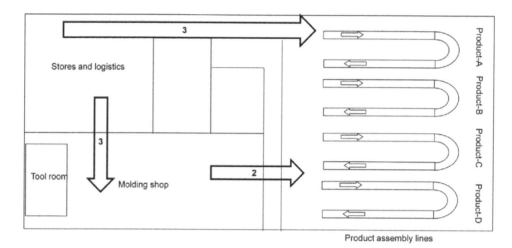

Figure 2.6 Inventory heads and material flow in current plant layout.

1. Electronics components
2. Plastic assemblies and small plastic parts from suppliers
3. Metallic subparts

Material flow from stores to the molding shop floor

1. Plastic granules
2. Plastic child parts for assembly on in-house molded plastic parts
3. Metallic child parts for assembly on in-house molded plastic parts

Material flow from molding shop to product assembly shop floor

1. Assembled plastic parts
2. Non-assembled plastic parts

In the current scenario of plant layout, a total of eight types of inventory heads need to be handled. That increases the **inventory handling cost** of the plant as well as increases the total **WIP material** of the plant.

Solution

If we implement the theory of grouping of work as per process dependent layout, we can move the assembly process done in the molding shop to integrated assembly lines for the respective products.

In this case, two inventories of raw materials of the molding shop as small plastic components and small metallic components will dissolve and one inventory of finished goods of the molding shop as assembled plastic parts will dissolve. The rest of the inventories will remain the same.

Proposed scenario

In the proposed scenario of plant layout, inventory heads will be five, as shown in Figure 2.7.

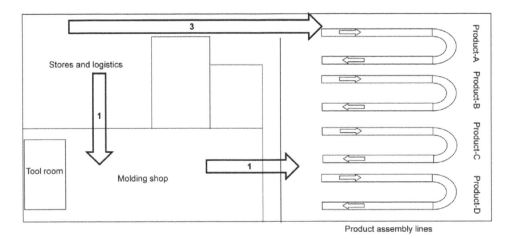

Figure 2.7 Inventory heads and material flow in proposed plant layout.

Material flow from store to product assembly shop floor

1. Electronics components
2. Plastic assemblies and small plastic parts from suppliers
3. Metallic subparts

Material flow from stores to the molding shop floor

1. Plastic granules

Material flow from molding shop to product assembly shop floor

1. Molded plastic parts

With the proposed solution, improvement in material flow and inventory handling will be

1. Inventory head will decrease to 5 from 8 (27.5% improvement in inventory handling).
2. Material flow will improve.
3. WIP will decrease.

2.3.2.2 Factor 2 – productivity

The process flow of the molding shop is shown in Figure 2.8. In the molding shop, small assembly lines (two or three assembly stations) near injection molding machines are there. As shown in the process flow chart in Figure 2.8, two times visual inspection is done for 100% parts.

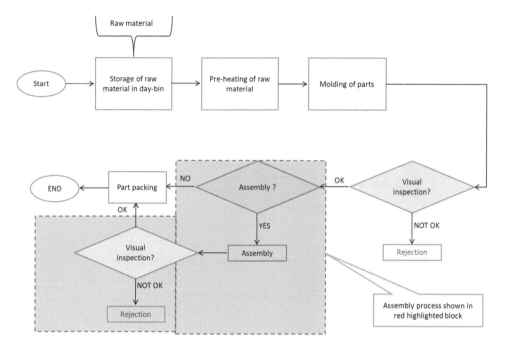

Figure 2.8 Process flow chart in molding shop.

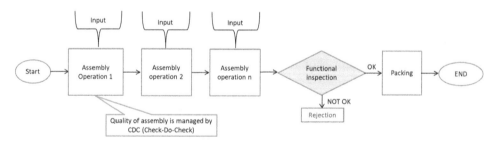

Figure 2.9 Process flow chart of product assembly line.

In Figure 2.9, a typical layout of the product assembly line is shown. Product assembly lines are designed in a U-shape. The quality of work of each station is managed by CDC (Check-Do-Check) process at each station. In the end, functional testing of the integrated product for the required functionality of components is done. All OK parts go for packing and Not OK in rejection.

Gaps in the current scenario in terms of work content management

1. Non-value-adding activity of visual inspection is done twice in the molding, once for the plastic part quality check after injection molding process and again after assembly for quality check of assembled part.

2. Bottleneck in the molding shop is the injection molding process cycle time (i.e., the cycle time of the molding machine). Hence, balancing the assembly line after it is very difficult which results in productivity loss in terms of idle time of manpower.

Solution

If we implement the theory of grouping of work as per process dependent layout, we can move the assembly process done in the molding shop to integrated assembly lines for the respective products.

In this case, we can save unnecessary inspections at two places (i.e., in the molding shop and on the product assembly shop floor for the assembled part). It will be easy and flexible to balance line and manage work content on the assembly shop floor for assembly processes.

Proposed scenario

For shifting the assembly on molded plastic parts process to the product assembly line, we first did a micro-level work content study. For that, we prepared a format, in which time study of small steps during assembly of child part on molded plastic parts is analyzed.

For example, as shown in Figure 2.10, it is the work content of an assembly workstation in the molding shop. From this study, we can assign some activities to another workstation where idle time is available while shifting work to the product assembly line from the molding shop. This will provide us with a better balancing % of the assembly line and higher productivity.

For example, while shifting this work to the product assembly line, kaizens were done and parallel doable activities were identified. Hence, a better sequence of work content in small studies has been done as shown in Figure 2.11.

Total numbers of assemblies were identified for each product assembly line and shifted to their respective product assembly lines. Total assembly operations in the molding shop are 20 (Figure 2.12).

After complete work-study, the entire assembly is shifted to the respective product assembly lines. Work content is divided and lines are balanced.

As shown in Figures 2.13–2.15, productivity improvement is done on every product assembly line with an average of **49% in product A, 100% in product B and 67% in product C.**

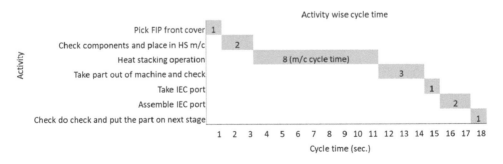

Figure 2.10 Micro work content study for assembly process.

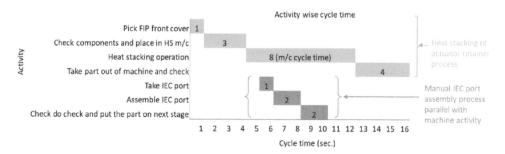

Figure 2.11 Improved micro work content for assembly process.

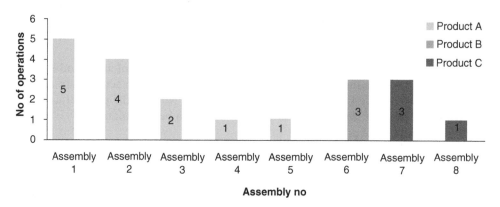

Figure 2.12 Total assembly operations shifting product family-wise.

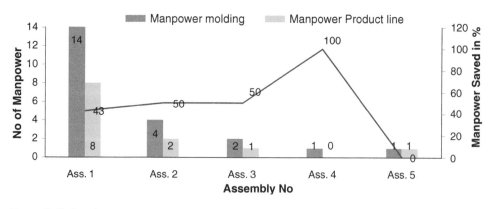

Figure 2.13 Productivity improvement in product A in each assembly.

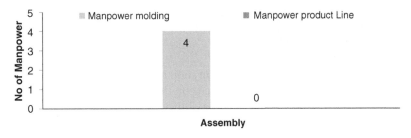

Figure 2.14 Productivity improvement in product B in each assembly.

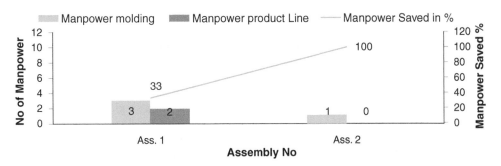

Figure 2.15 Productivity improvement in product C in each assembly.

2.3.2.3   Factor 3 – quality

Quality is the most important parameter in manufacturing. In the current scenario, 90% of problems reported by assembly lines to the molding shop are related to assembly parts.

Analysis of this is done and found that supervisors and line leaders are specialists in the injection molding process and are very poor in handling assembly processes. Since the injection molding process is a vast technology of manufacturing, it required an especially skilled person for this technology to handle the shop floor. Also, manpower skill is one of the important factors for quality. People on the assembly line do have skills of assembly, but the manpower of the molding shop should have both skills of injection molding and assembly.

Gaps in the current scenario

1. Manpower skill counts are increased due to dual operation.
2. Staff skill counts are increased due to dual operation.
3. Flexibility in manpower deployment is less due to skill limitations.

Solution

If we implement the theory of grouping of work as per process dependent layout, we can move the assembly process done in the molding shop to integrated assembly lines for the respective products.

Proposed scenario

In the proposed scenario, the molding shop is completely de-skilled for the assembly processes and now manpower can focus on the single aim of improving the quality of injection-molded plastic parts. The flexibility of manpower deployment is high in the molding shop now.

2.3.2.4   Factor 4 – safety

Safety in a plant means the safety of people, equipment and material. In the current layout, the molding shop has space constraints for movement. Molding is a heavy machinery work area, where frequent mold changeover happens. In such conditions, safety becomes a crucial area to improve. In heavy machinery areas, congested and busy passages are dangerous.

In the current scenario of layout, in the molding shop, mold movement with Crain, material and manpower movement, all are happening from a single channel of gangway, as shown in Figure 2.16.

After implementation of the new methodology and assembly transfer to the product assembly line, the area occupied with assembly child part and assembled parts is used for movement of material and manpower from the backside of machines. In the improved layout of the molding shop, Gangway is only being used for Crain movement. This scenario creates a better and safer situation in the molding shop (Figure 2.17).

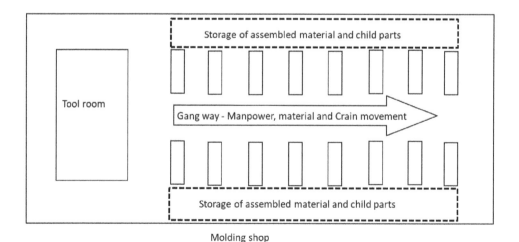

Molding shop

Figure 2.16 Layout of molding shop for movement.

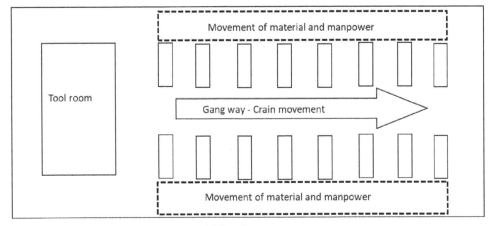

Molding shop

Figure 2.17 Improved layout of molding shop for movement.

2.4   DISCUSSION AND RECOMMENDATIONS

Implementation of new designs and improvements are the most important factor behind the success of a company. Change is constant and there is always a better way of doing things. Everyone in the team from the top management to the co-workers should support and adopt the changes. In the optimized facility layout, saving in terms of less inventory handling and productivity increase happened and the quality of products improved. Further optimization is possible when new production lines will be installed in the plant. For future products, during the Design for Manufacturability (DFM) of the product line, this project should be considered and parts should be designed with no post-molding operations in the molding shop. Safety should be a considerable point while designing a layout. The layout should be modified timely to improve the safety conditions of the plant. A new method of understanding work content of assembly processes is visible in this study in the form of micro work content study. It is a useful tool for line balancing and can be used in future for line balancing purposes. The facility layout should be visited by management timely and factors should be examined with respect to the layout. Observation-based improvements can be carried out; hence, ideas of everyone should be grown and respected by the company.

Key findings: This study indicates that layout problems are not always complex and don't always require complex mathematics to work upon layouts. Time-to-time review of assigned spaces and works can also improve hugely in layouts. Layout significantly affects operations and has a huge work of scope every time.

Limitations: This study is a case study and limited to the on-site work and a specific type of plant layout and depends on the opportunity availability of changes.

Future research scope: Methodology can be modified for the measurement of work of scope in layout for any type of designed layout against all four factors included in this study.

2.5 CONCLUSION

The improved layout with this change approach lead to better flow of material within the plant, reduced material handling, increased productivity for certain operations and improved quality and safety conditions in heavy machinery areas. The facility layout proposed in this chapter is inclined toward a hybrid layout of both process- and product-based approaches as integrated assembly lines are there for a variety of product assembly and molding and assembly processes are distinguished in two different areas. The observation in the current layout and study resulted in a 27.5% reduction in inventory handling, 72% productivity improvement for certain operations and significant improvement in the quality and safety of the plant. Re-layout is a very effective tool for continuous process improvement in manufacturing plants. During this study, various more alternatives for re-layout were also tested with respect to these four variables (Inventory, Productivity, Quality and Safety) and the final changes were done. On the behalf of this case study, it can be concluded that these four factors were significantly improved in the plant by re-layout.

REFERENCES

Amar, S.H., Abouabdellah, A., 2016, Facility layout planning problem: effectiveness and reliability evaluation system layout designs. *IEEE, International Conference on System Reliability and Science*, Paris, pp. 110–114.

Balakrishnan, J., Cheng, C.H., 2007, Multi-period planning and uncertainty issues in cellular manufacturing: a review and future directions. *European Journal of Operational Research*, 177:281–309. doi: 10.1016/j.ejor.2005.08.027.

Barnwal, S., Dharmadhikari, P., 2016, Optimization of plant layout using SLP method, *IJIRSET*, 5(3):3008–3015, http://www.ijirset.com/upload/2016/march/46_Optimization.pdf.

Chandra Mouli, C. S. B., Kulkarni, S. D., Deepak, S., 2021, Productivity improvement of a small scale industry by the application of an effective plant layout and weld-fixture design, *Materials Today: Proceedings*, ISSN 2214-7853, doi: 10.1016/j.matpr.2021.09.060. https://www.sciencedirect.com/science/article/pii/S2214785321058569.

Chen, C., et al., 2019, Optimal facility layout planning for AGV-based modular prefabricated manufacturing system, *Automation in Construction*, 98:310–321.

Kovács, G., Kot, S., 2017, Facility layout redesign for efficiency improvement and cost reduction, *Journal of Applied Mathematics and Computational Mechanics* 16(1):63–74.

Meller, R. D., Gau, K.-Y., 1996, The facility layout problem: recent and emerging trends and perspectives, *Journal of Manufacturing Systems*, 15(5):351–366, ISSN 02786125, doi: 10.1016/0278-6125(96)84198-7. https://www.sciencedirect.com/science/article/pii/0278612596841987.

Nyemba, W. R., Mbohwa, C., 2017, Process mapping and optimization of the process flows of a furniture manufacturing company in Zimbabwe using machine distance matrices, *Procedia Manufacturing*, 8:447–454, ISSN 2351-9789, doi: 10.1016/j.promfg.2017.02.057. https://www.sciencedirect.com/science/article/pii/S235197891730063X.

Schenk, M., Wirth, S., Müller, E., 2014, Fabrikplanung und Fabrikbetrieb: Methoden Für die Wandlungsfähige, Vernetzte und Ressourceneffziente Fabrik. VDI-Buch, 2. vollst. überarb. u. erw. Aufl., Berlin. Springer Berlin Heidelberg, Heidelberg.

Tarigan, U., et al., 2019, Facility layout design through integration of lean manufacturing method and CORELAP algorithm in concrete factory, *IOP Conference Series: Materials Science and Engineering*, 505:1–7.

Tarigan, U., Herdita, G. Y., Tarigan, U. P. P., 2019, Redesign of facility layout with graph method and genetic algorithm in wood manufacturing plant, *Journal of Physics: Conference Series*, 1230:1–8.

Wu, Y., Wang, Y., 2017, A chemical industry area-wide layout design methodology for piping implementation, *Chemical Engineering Research and Design*, 118:81–93, ISSN 02638762, doi: 10.1016/j.cherd.2016.12.005. https://www.sciencedirect.com/science/article/pii/S02638762 16304749.

Zhang, M., Batta, R., Nagi, R., 2009, Designing manufacturing facility layouts to mitigate congestion, http://www.acsu.buffalo.edu/~batta/designing.pdf, downloaded: 02.09.2016.

Chapter 3

Multi-criteria decision analysis applications and trends in manufacturing domain

Anita Kumari and Bappa Acherjee

CONTENTS

3.1   INTRODUCTION

Rapidly evolving product transformations have necessitated a similarly fast reaction from the manufacturing industry. Manufacturing industries must choose suitable strategic options, product specifications, production processes, work materials and tools, as well as plant and equipment, to address these difficulties. Strategic decisions are complicated since decision-making is becoming increasingly difficult. Manufacturing sectors must make logical decisions due to the large levels of investments and the inherently uncertain markets. As a result, decision-making is critical in strategic planning for businesses. Decision-making entails defining and selecting options depending on the decision-maker's values, preferences, and perceptions. Several dynamic process stages are included in decision-making, such as choosing parameters, assessing alternatives, collecting and reviewing processing information, generating and evaluating incomplete or intermediate outcomes, reconsidering the criteria, alternatives, and information based on the achieved result, and continuing the process before an actionable outcome is achieved (Onut et al., 2008; Zavadskas et al., 2016a).

MCDM focuses on aiding decision-makers who are confronted with multiple and contradictory options in making the right choice (Emovon and Oghenenyerovwho, 2020). The MCDM approaches are used in a variety of areas, including medicine, finance and banking, environmental management, community design, robotics, energy planning, nuclear disaster management, equipment selection, infrastructure, manufacturing, etc. (Zavadskas et al., 2016a). The following are among the most significant manufacturing decision-making scenarios:

- Choosing the right material for a specific structural segment,
- Appraisal of best product designs,
- Assortment and evaluation of advanced machining processes,
- Assortment and appraisal of flexible manufacturing systems,

DOI: 10.1201/9781003293576-3

- Cutting fluid selection for a specific machining operation,
- Appraisal of machinability of work materials,
- Choosing industrial robots for specific manufacturing applications,
- Identification of the root cause of machine tool failure,
- Choosing material handling equipment,
- Choosing automated inspection equipment, etc.

The identification of appropriate material for any product is among the most critical and difficult jobs in a manufacturing organisation. The right material selection for engineering components presents numerous benefits, such as enhanced durability and performance, cost reduction, longer life cycle, etc., whereas the wrong choice of material results in poor functionality and eventual quality problem (Chatterjee et al., 2018). The selection of material may be thought of as an MCDM challenge, requiring a rational and rigorous material selection approach to determine the best option (Maity et al., 2012). The material selection procedure is complicated owing to the large variety of available substitute materials and their various interactions with various eligibility requirements. MCDM models assist in the realisation of engineering goals by using some form of attribute weighting (Tian et al., 2018). The major purpose of material selection is to maintain prices in check while still attaining product performance targets. In aircraft industries, for example, weight minimisation is among the primary priorities for design upgrades while maintaining a close eye on the component's cost, performance, and efficiency.

Several manufacturing methods are used in the production of various products. Not all manufacturing processes can manufacture a product with the same ease, accuracy, and cost-effectiveness. Each manufacturing process usually has some advantages and disadvantages over the others. Advanced manufacturing processes employ sophisticated and advanced computerised systems, each with its own distinct set of prerequisites and advantages (Prasad and Chakraborty, 2014). Engineers face a variety of discrete process selection issues in the manufacturing sectors. A variety of MCDM approaches have been proposed to aid policymakers in solving such types of optimisation issues. The MCDM approaches can assist in determining the optimal solution for making the production process more cost-effective (Karande and Chakraborty, 2012; Madic et al., 2015). The present paper surveys the application trends in the use of MCDM approaches to manufacturing decision analysis, with a focus on the selection of material and manufacturing processes for manufacturing applications.

3.2   APPLICATIONS AND TRENDS OF MCDM IN MANUFACTURING DECISION-MAKING

Several academics have proposed MCDM approaches to address the problem of selecting suitable materials and manufacturing processes. The primary goal of the MCDM-based selection strategy is to find the right set of criteria for a specific purpose, specify the specifics associated with these criteria, and establish methods to evaluate these criteria to satisfy the final requisites. Thus, in order to fulfil the needs of the manufacturing sectors, researchers have worked diligently to provide effective methodologies for selecting viable alternatives for a variety of engineering products. MCDM acts as a bridge between decision-makers and experts, directing them to a decision when various and often overlapping factors are involved. Table 3.1 summarises the historical evolution and operational methodologies of the well-known MCDM methods.

In many situations, one MCDM technique is combined with another MCDM method or with additional methods, like fuzzy sets or grey numbers, to assess the comparative significance of criteria, resulting in hybrid MCDM (HMCDM) methods (Zavadskas et al., 2016a). The key concept of HMCDM methods is depicted in Figure 3.1. Several limitations of classical MCDM approaches can be addressed by using HMCDM methods. The use of more than one MCDM approach aids in the integration of results in terms of final decision-making. The hybrid method addresses several challenges, such as deciding parameter values and weights and translating those into multi-attribute utility function values (Ayag, 2007).

Models for decision-making must be as realistic as feasible. Fuzziness in decision-making also arises from a sense of managerial confusion, in which inconsistencies and complexities make obtaining an appropriate choice challenging. As a result, combining MCDM with grey numbers or fuzzy sets is favoured. Fuzzy logic can be useful in overcoming inconsistencies caused by human contextual judgements and imperfect choice relationships (Taha and Rostam, 2011). Other methods may also be used to provide additional justification to the problem formulation. Quantitative and qualitative analytical approaches can be used to generalise knowledge, pick sustainable appraisal metrics, and derive evaluation criteria for multiple criteria analysis (Zavadskas et al., 2016a).

Roth et al. (1994) used MAUA (Multi-Attribute Utility Analysis) to assess the efficiency of several camshaft materials for use in vehicle camshafts, taking into account decision variables such as material modulus, compressive yield strength, weight, and cost. Ayag (2007) proposed an HMCDM method for selecting machine tools for specific

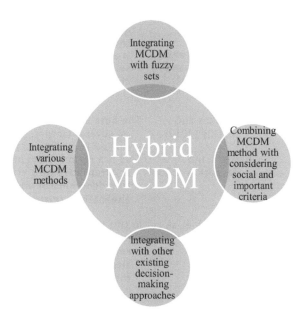

Figure 3.1 The general structure of hybrid MCDM methods.

Table 3.1 Historical evolution and functional features of the well-known MCDM techniques

MCDM methods	Developed by	Operational methodologies
ELECTRE ("ELimination et choix traduisant la realité")	Roy (1968)	The evaluation is based on partial preference aggregation and can cope with qualitative data with significant uncertainty.
SAW (Simple Additive Weighting)	MacCrimmon and Rand (1968)	The fundamental idea is to calculate the weighted total of performance ratings for each option across all characteristics.
DEMATEL (Decision-Making Trial and Evaluation Laboratory)	Fontela and Gabus (1976)	It is concerned with assessing interconnected relations between criteria and identifying the essential ones using a visual structural model.
DEA (Data Envelopment Analysis)	Charnes et al. (1978)	A linear programming approach for comparing the relative efficiency of alternatives.
AHP (Analytic Hierarchy Process)	Saaty (1980)	Local priorities or weighting factors are determined by comparing criteria and sub-criteria pair-wise.
PROMETHEE (Preference Ranking Organisation Method for Enrichment Evaluations)	Mareschal et al. (1984)	It is an outranking approach for rating a finite collection of alternatives by scrutinising their conflicts and complementarity, as well as highlighting the best alternative and the organised rationale behind this.
TOPSIS (Technique of Order Preference Similarity to the Ideal Solution)	Hwang and Yoon (1981)	The preferred alternative is chosen based on the choice that is the furthest away from the negative-ideal response and the closest to the positive-ideal response.
COPRAS (Complex Proportional Assessment)	Zavadskas et al. (1994)	This approach employs a sequential ranking and evaluation of alternatives based on their importance and utility.
VIKOR ("VIseKriterijumska optimizacija i komoromisno resenje")	Opricovic (1998); Opricovic and Tzeng (2002)	It is a compromise ranking method for selecting the finest option that relates closely to the positive-ideal response.
ARAS (Additive Ratio Assessment)	Zavadskas and Turskis (2010)	The best alternative is selected depending on several factors, and alternatives are ranked according to their level of utility.
MOORA (Multi-objective Optimisation on the basis of Ratio Analysis)	Brauers and Zavadskas (2006)	This is a ratio system in which each value of an alternative on a criterion is compared to a denominator which represents all alternatives related to that criterion.
SWARA (Step-wise Weight Assessment Ratio Analysis)	Keršuliene et al. (2010)	In this technique, the decision-maker's judgement determines the relative relevance and initial ranking of alternatives for each characteristic, and then the weighting of each characteristic is calculated.

(Continued)

Table 3.1 (Continued) Historical evolution and functional features of the well-known MCDM techniques

MCDM methods	Developed by	Operational methodologies
WASPAS (Weighted Aggregated Sum Product Assessment)	Zavadskas et al. (2012)	This is a one-of-a-kind combination of two methods: WSM (Weighted Sum Method) and WPM (Weighted Product Method).
CODAS (Combinative Distance-based Assessment)	Zavadskas et al. (2016b)	This method assesses the desirability of an alternative using the Euclidean and Taxicab distances, respectively, as the primary and secondary metrics, and both distances are measured with respect to the negative-ideal solutions.

applications by combining AHP with simulation methodology. Onut et al. (2008) selected and evaluated appropriate machine tools for vertical CNC (Computer Numerical Control) machining using a hybrid of fuzzy-AHP and fuzzy-TOPSIS approaches. Chatterjee et al. (2009) utilised the ELECTRE and VIKOR techniques for flywheel and boat mast material selection. Yurdakul and Tansel (2009) used Spearman's rank correlation coefficient in conjunction with fuzzy-TOPSIS approach to address the complexity and uncertainty issues related to the machine tool selection tasks. Das and Chakraborty (2011) utilised ANP (Analytic Network Process) for determining the finest unconventional machining method for certain machining purpose. Taha and Rostam (2011) employed fuzzy-AHP and fuzzy-ANN ("Artificial Neural Network") hybrid approaches to evaluate machine tools to be used in a flexible manufacturing system. Chatterjee et al. (2011) utilised the COPRAS and EVAMIX (Evaluation of Mixed Data) approaches for selecting cryogenic storage tank material for very-low temperature use. Sadhu and Chakraborty (2011) utilised DEA (Data Envelopment Analysis) for selecting the unconventional machining system for generating specific shape features on a chosen work material. Maity et al. (2012) used the COPRAS-G (COPRAS with Grey Numbers) method to choose the cutting tool material. Chauhan and Vaish (2012) implemented VIKOR and TOPSIS techniques for selecting hard and soft magnetic materials. To quantify weights of various characteristics, Shannon's entropy approach is used, and the clustering method is employed for categorising magnetic materials. Girubha and Vinodh (2012) investigated the efficacy of the fuzzy-integrated VIKOR technique for material selection in vehicle parts.

Karande and Chakraborty (2012) utilised an approach combining PROMETHEE and GAIA (Geometrical Analysis for Interactive Aid) for choosing the appropriate non-conventional method, which would be useful for process engineers as a visual decision aid. Ilangkumaran et al. (2013) used a hybrid strategy that included fuzzy-AHP, PROMETHEE-I, and PROMETHEE-II approaches to find the optimum material to use in the production of an automotive bumper. The criterion weights are computed using fuzzy-AHP, the departing and entering flows are determined using PROMETHEE-I, and the overall score of the material is determined using PROMETHEE-II. Caliskan et al. (2013) employed EXPROM-II ("EXtension of PROMETHEE-II"), TOPSIS, and VIKOR approaches to identify suitable hard milling tool holder material. Chatterjee and Chakraborty (2013) utilised COPRAS and ARAS techniques to address the material selection problem of gear in a specified manufacturing context. Maity and Chakraborty (2013) adopted a fuzzy-TOPSIS approach for deciding on abrasives for grinding wheels. Jajimoggala and Karri (2013) utilised a hybrid Fuzzy-AHP-TOPSIS strategy for identifying the best impellor material, and AISI4340 is determined as being the material of choice among different options. Cavallini et al. (2013) devised a method for determining the best coating for reducing abrasion on an Al7075 aluminium alloy utilising QFD (Quality Function Deployment) and the VIKOR approaches. Mansor et al. (2014) utilised a hybrid AHP-TOPSIS technique to choose a thermoplastic polymer for designing an automobile brake component for a passenger car. Rai et al. (2013) used and contrasted VIKOR and regret theory-based VIKOR to investigate a scenario of flywheel material selection. The fracture toughness, fatigue limit, fragmentability, and cost are the decision factors utilised for flywheel material selection. Prasad and Chakraborty (2014) employed the QFD method for selecting the unconventional process for standard holes in superalloys, standard through cavities in ceramics, deep through cutting in titanium, and double contouring in duralumin. Liu et al. (2014) used a combination of DANP (DEMATEL-based ANP) and modified-VIKOR for split journal bearing material selection.

Anojkumar et al. (2014) conducted a case study on pipe material shortlisting for use in the sugar industry using four distinct MCDM approaches that merged fuzzy-AHP with PROMETHEE, ELECTRE, VIKOR, and TOPSIS with the goal of conducting a comparative study of MCDM approaches. Nguyen et al. (2014) used fuzzy-ANP in conjunction with the COPRAS-G approach to identify the suitable machine tool, and the results were compared against TOPSIS-G (TOPSIS with Grey Numbers), SAW-G (SAW with Grey Numbers), and GRA (Grey Relational Analysis). Yazdani and Payam (2015) chose the optimal material for MEMS ("Micro-Electro-Mechanical System") using the Ashby method, as well as the TOPSIS and VIKOR approaches. Madic et al. (2015) suggested a technique incorporating AHP, MOORA, and TOPSIS for selecting the most effective unconventional machining processes considering material use and various performance requirements. Govindan et al. (2015) used a hybrid DANP and PROMETHEE to evaluate green manufacturing practices for rubber tyre and tube manufacturers. Yuvaraj and Kumar (2015) optimised AWJM (Abrasive Water Jet Machining) settings using the TOPSIS method to improve product quality. The TOPSIS method is utilised (by Sahu and Sahu, 2015) to identify the finest CNC machine tool. Yazdani et al. (2016) implemented SWARA, WASPAS, and MOORA for evaluating low electrical resistivity materials for MEMS and hard magnetic materials. Each approach is used independently, and the results are compared to ensure accuracy. Li et al. (2016) utilised the AHP and entropy techniques to choose cutting tool material, with the AHP technique subjectively assessing the relevance of material qualities and assigning criteria weights to them, whereas the entropy-based approach objectively assesses the importance and assigns weighting factors. Kumar and Ray (2016) employed entropy and TOPSIS approaches for selecting nitride steel material for engineering design, in which the entropy method is applied to compute weight factors while the TOPSIS method is applied for ranking of alternatives. Madic et al. (2016) utilised AHP, OCRA ("Operational Competitiveness RAting"), and WASPAS to identify the optimum laser cutting manufacturing process conditions. Zhao et al. (2016) developed a method for grading prospective sustainable design materials, which combines GRA and AHP.

Sen et al. (2016) employed VIKOR, ARAS, TOPSIS, MOORA, and COPRAS to pick the best connecting rod material, analysed the efficiency of the alternatives using multiple decision criteria, and made a comparative review of the methods used. Nasab and Anvari (2017) adopted a materials selection strategy based on TOPSIS, COPRAS, and DEA, in which TOPSIS and COPRAS are applied to assess alternatives and DEA is utilised as a rough guideline for selection. Shukla et al. (2017) surveyed the use of the TOPSIS method in a variety of manufacturing applications, including turning, drilling, milling, abrasive jet machining, electric discharge machining, micromachining, and other specialised forms of machining. Chatterjee et al. (2018) suggested a hybrid technique utilising DoE (Design of Experiments) and EDAS (Evaluation based on Distance from Average Solution) methodologies for selecting the optimum gear material for high-stress and high-speed operational conditions, with carburised steel selected as the best choice. Zaman et al. (2018) used the AHP method for picking the best material and procedure for advanced additive manufacturing applications in aerospace industries. Tian et al. (2018) proved the efficacy of the AHP, GC-TOPSIS technique for determining the final rating of green decorating materials and selecting the best option. To validate the proposed technique, 10 various types of solid woods are utilised in a case study, which assists in dealing with the shortcomings of a single method.

Gul et al. (2018) used the fuzzy-PROMETHEE approach to decide the correct material for a vehicle instrument console, and the results are contrasted with those produced by fuzzy-ELECTRE, fuzzy-TOPSIS, and fuzzy-VIKOR. Prasad and Chakraborty (2018) created a decision-making framework by implementing PROMETHEE, VIKOR, and TOPSIS methods in Visual Basic 6.0 for selecting appropriate unconventional machining processes for specified materials and form features. Patnaik et al. (2019) implemented AHP-TOPSIS approach to pick the suitable wear resistance material from the standpoint of structural applications. Singh et al. (2019) utilised TOPSIS and ARAS methods to determine the ultrasonic machining conditions for obtaining optimum hole quality on a specified composite, and the best alternative was evaluated. Buyukozkan and Gocer (2019) suggested the Pythagorean fuzzy set and fuzzy-CODAS for the evaluation of additive manufacturing technologies. Moradian et al. (2019) contrasted the AHP-based VIKOR, TOPSIS, and MOORA techniques in valve body material selection of a brake system based on several factors. Niu et al. (2020) employed RBR (Rule-based Reasoning) and COPRAS-G in the cutting tool material selection problem, with RBR being used to optimise the set of options as well as the computation intricacy of the COPRAS-G technique. Raigar et al. (2020) suggested a hybrid approach based on PIV (Proximity-Indexed Value) and BWM (Best-Worst Method) to identify the best additive manufacturing technology for spur gear production. The material jetting technique, among the various additive manufacturing technologies, provides the most dimensionally precise as well as superior-quality components. Lo et al. (2020) presented a hybrid DEMATEL and TOPSIS risk assessment model that combines various methodologies to provide an FMEA (Failure Modes and Effects Analysis) model for the creation of exhaustive failure-mode rating. Patel et al. (2020) suggested a hybrid MOORA, GRA, TOPSIS, and DEAR (Data Envelopment Analysis-Based Ranking) method for improving AWJM for polymer matrix composites. Ibrahim et al. (2020) used a combination of TOPSIS, VIKOR, and ANOVA (Analysis of Variance) to assess the efficiency of the submerged abrasive water jet turning process to improve the machinability of cast amide. Dohale et al. (2021) used AHP to pick the best possible additive manufacturing technique among those available based on a set of desired parameters. Rao and Lakshmi (2021) proposed a new MADM ("Multiple Attribute Decision Making") method named as R-method and applied it for vendor selection, industrial robot selection, material selection for industrial products, and flexible manufacturing system selection. Kumari and Acherjee (2021) applied the CRITIC ("CRiteria Importance Through Intercriteria Correlation") -CODAS ("COmbinative Distance-based Assessment") method to identify a non-conventional machining process for titanium processing, and PAM (Plasma Arc Machining) is determined to be the best option. Leong et al. (2022) applied a hybrid GRA-BWM-TOPSIS to evaluate resilient supplier selection when there are uncertainties and incomplete data.

3.3 CONCLUSION

MCDM methods are commonly utilised in the manufacturing industry to aid decision-making via alternative assessment and comparative evaluation. The most significant advantage of MCDM techniques is their capacity to solve issues with many competing interests. According to the literature review, decision-making using MCDM techniques is driven by a range of variables such as the set of alternatives and attributes, the agility of the decision-making process, computational intricacy, the inclusion or elimination of a criterion, efficacy in facilitating a collective decision, and so on. MCDM techniques are intended to identify a preferred alternative, categorise alternatives into a particular class, and grade alternatives in subjective order of preference. It is found that various MCDM methods produce different rankings, and therefore, choosing a suitable method is a difficult task. It is thus important to employ a hybrid strategy based on multiple approaches and to combine the advantageous features of each for the final decision. The capacity to combine subjective and quantitative elements into the value of utility function is what distinguishes hybrid techniques from individual techniques. Simultaneous use of fuzzy logic can help resolve ambiguities caused by human contextual judgements and imperfect preference correlations, and bring a model closest to realistic approximation.

The application of HMCDM approaches to complex issues is growing significantly because of their capacity to competently aid decision-makers in coping with a diverse set of facts. HMCDM methods are now widely used and provide better results in unpredictable situations. There is an opportunity to showcase HMCDM techniques that are simple in structure and simple to implement in manufacturing decision-making. There are opportunities to evaluate several MCDM techniques to select the best one for assessing and selecting materials and manufacturing processes for enhanced product- -performance.

REFERENCES

Anojkumar, L., Ilangkumaran, M., & Sasirekha, V. (2014). Comparative analysis of MCDM methods for pipe material selection in sugar industry. *Expert Systems with Applications*, 41, 2964–2980.
Ayag, Z. (2007). A hybrid approach to machine-tool selection through AHP and simulation. *International Journal of Production Research*, 45, 2029–2050.

Brauers, W., & Zavadskas, E. K. (2006). The MOORA method and its application to privatization in a transition economy. *Control and Cybernetics*, 35, 445–469.

Buyukozkan, G., & Gocer, F. (2019). Assessment of additive manufacturing technology by Pythagorean fuzzy CODAS. *International Conference on Intelligent and Fuzzy System*, 1029, 959–968.

Caliskan, H., Kursuncu, B., Kurbanoglu, C., & Guven, S. Y. (2013). Material selection for the tool holder working under hard milling conditions using different multi criteria decision making methods. *Materials and Design*, 45, 473–479.

Cavallini, C., Giorgetti, A., Citti, P., & Nicolaie, F. (2013). Integral aided method for material selection based on quality function deployment and comprehensive VIKOR algorithm. *Materials and Design*, 47, 27–34.

Charnes, A., Cooper, W. W, & Rhodes, E. (1978). Measuring the efficiency of decision-making units. *European Journal of Operational Research*, 2, 429–444.

Chatterjee, P., Athawale, V. M., & Chakraborty, S. (2009). Selection of materials using compromise ranking and outranking methods. *Materials and Design*, 30, 4043–4053.

Chatterjee, P., Athawale, V. M., & Chakraborty, S. (2011). Materials selection using complex proportional assessment and evaluation of mixed data methods. *Materials and Design*, 32, 851–860.

Chatterjee, P., & Chakraborty, S. (2013). Gear material selection using complex proportional assessment and additive ratio assessment-based approaches: A comparative study. *International Journal of Materials Science and Engineering*, 1, 104–111.

Chatterjee, P., Banerjee, A., Mondal, S., Boral, S., & Chakraborty, S. (2018). Development of a hybrid meta-model for material selection using design of experiment and EDAS method. *Engineering Transactions*, 66, 187–207.

Chauhan, A., & Vaish, R. (2012). Magnetic material selection using multiple attribute decision making approach. *Materials and Design*, 36, 1–5.

Das, S. & Chakraborty, S. (2011). Selection of non-traditional machining processes using analytic network process. *Journal of Manufacturing Systems*, 30, 41–53.

Dohale, V., Akarte, M., Gupta, S., & Verma, V. (2021). Additive manufacturing process selection using MCDM, *Advances in Mechanical Engineering*. Lecture Notes in Mechanical Engineering, pp. 601–609, Springer, Singapore.

Emovon, I., & Oghenenyerovwho, O. S. (2020). Application of MCDM method in material selection for optimal design: A review. *Results in Materials*, 7, 100–115.

Fontela, E., & Gabus, A. (1976). The DEMATEL observer, DEMATEL 1976 Report. Switzerland Geneva: Battelle Geneva Research Center, USA.

Girubha, R. J., & Vinodh, S. (2012). Application of fuzzy VIKOR and environmental impact analysis for material selection of an automotive component. *Materials and Design*, 37, 478–486.

Govindan, K., Kannan, D., & Shankar, M. (2015). Evaluation of green manufacturing practices using a hybrid MCDM model combining DANP with PROMETHEE. *International Journal of Production Research*, 53, 6344–6371.

Gul, M., Celik, E., Gumus, A. T., & Guneri, A. F. (2018). A fuzzy logic based PROMETHEE method for material selection problems. *Beni-Suef University Journal of Basic and Applied Sciences*, 7, 68–79.

Hwang, C., & Yoon, K. (1981). *Multiple Attribute Decision Making: Methods and Applications, A State of the Art Survey*, New York: Sprinnger-Verlag.

Ibrahim, S. A. B., Korkmaz, S., Cetin, M. H., & Kartal, F. (2020). Performance evaluation of the submerged abrasive water jet turning process for improving machinability of cast amide. *Engineering Science and Technology, An International Journal*, 23, 801–811.

Ilangkumaran, M., Avenash, A., Balakrishnan, V., Kumar, S. B., & Raja, M. B. (2013). Material selection using hybrid MCDM approach for automobile bumper. *International Journal of Industrial and Systems Engineering*, 14, 20–39.

Jajimoggala, S. & Karri, R. R. (2013). Decision making model for material selection using a hybrid MCDM technique. *International Journal of Applied Decision Science*, 6, 144–159.

Karande, P. & Chakraborty, S. (2012). Application of PROMETHEE-GAIA method for non-traditional machining process selection. *Management Science Letters*, 2, 2049–2060.

Keršuliene, V., Zavadskas, E. K., & Turskis, Z. (2010). Selection of rational dispute resolution method by applying new step-wise weight assessment ratio analysis (Swara). *Journal of Business Economics and Management*, 11, 243–258.

Kumar, R., & Ray, A. (2016). Selection of material for optimal design using multi-criteria decision making. *Procedia Materials Science*, 6, 590–596.

Kumari, A., & Acherjee, B. (2021). Selection of non-conventional machining process using CRITIC-CODAS method. *Material Today: Proceedings*. In press.

Leong, W. Y., Wong, K. Y., & Wong, W. P. (2022). A new integrated multi-criteria decision-making model for resilient supplier selection. *Applied System Innovation*, 5, 1–18.

Li, A., Zhao, J., Gong, Z., & Lin, F. (2016). Optimal selection of cutting tool materials based on multi-criteria decision-making methods in machining Al-Si piston alloy. *The International Journal of Advanced Manufacturing Technology*, 86, 1055–1062.

Liu, H. C., You, J. K., Zhen, L., & Fan, X. J. (2014). A novel hybrid multiple criteria decision-making model for material selection with target-based criteria. *Material and Design*, 60, 380–390.

Lo, H. W., Shiue, W., Liou, J. J., & Tzeng, G. H. (2020). A hybrid MCDM-based FMEA model for identification of critical failure modes in manufacturing. *Soft Computing* 24, 15733–15745.

MacCrimmon, K. R., & Rand, C. (1968). *Decision Making Among Multiple-Attribute Alternatives: A Survey and Consolidated Approach*. Santa Monica, CA: Rand Corp.

Madic, M., Radovanovic, M., & Petkovic, D. (2015). Non-conventional machining process selection using multi-objective optimization on the basis of ratio analysis method. *Journal of Engineering Science and Technology*, 10, 1441–1452.

Madic, M., Antucheviciene, J., Radovanovic, M., & Petkovic, D. (2016). Determination of manufacturing process conditions by using MCDM methods: Application in laser cutting. *Inzinerine Ekonomika-Engineering Economics*, 27, 144–150.

Maity, S. R., Chatterjee, P., & Chakraborty, S. (2012). Cutting tool material selection using grey complex proportional assessment method. *Materials and Design*, 36, 372–378.

Maity, S.R., & Chakraborty, S. (2013). Grinding wheel abrasive material selection using fuzzy TOPSIS method. *Materials and Manufacturing Processes*, 28, 408–417.

Mansor, M. R., Sapuan, S. M., Syams, Zainudin, E. S., Nuraini, A. A., & Hambali, A. (2014). Application of integrated AHP-TOPSIS method in hybrid natural fiber composites materials selection for automotive parking brake lever component. *Australian Journal Basic and Applied Science*, 8, 431–439.

Mareschal, B., Brans, J. P., & Vincke, P. (1984). PROMETHEE: A new family of outranking methods in multicriteria analysis. ULB Institutional Repository, ULB–Universite Libre de Bruxelles, Brussels.

Moradian, M., Modanloo, V., & Aghaiee, S. (2019). Comparative analysis of multi criteria decision making techniques for material selection of brake booster valve body, *Journal of Traffic and Transportation Engineering*, 6, 526–534.

Nasab, S. M., & Anvari, A. S. (2017). A comprehensive MCDM-based approach using TOPSIS, COPRAS and DEA as an auxiliary tool for material selection problems. *Materials and Design*, 121, 237–253.

Niu, J., Huang, C., Li, C., Zou, B., Xu, L., Wang, J., & Liu, Z. (2020). A comprehensive method for selecting cutting tool materials, *The International Journal of Advanced Manufacturing Technology*, 110, 229–242.

Nguyen, H. T., Dawal, S.J., Nukman, Y., & Aoyama, H. (2014). A hybrid approach for fuzzy multi-attribute decision making in machine tool selection with consideration of the interactions of attributes. *Expert Systems with Applications*, 41, 3078–3090.

Onut, S., Kara, S. S., & Efendigil, T. (2008). A hybrid fuzzy MCDM approach to machine tool selection. *Journal of Intelligent Manufacturing*, 19, 443–453.

Opricovic, S. (1998). Multicriteria optimization of civil engineering systems. PhD Thesis, (Serbian) *Belgrade: Faculty of Civil Engineering*, 1–302.

Opricovic, S., & Tzeng, G. H. (2002). Multicriteria planning of post-earthquake sustainable reconstruction. *Computer-Aided Civil and Infrastructure Engineering*, 17, 211–220.

Patel, C. G. M., Kumar, R. S., & Naidu, N. V. S. (2020). Optimization of abrasive water jet machining for green composites using multi-variant hybrid techniques. In: Gupta, K., Gupta, M. (eds.), *Optimization of Manufacturing Processes*, 129–162, Springer Series in Advanced Manufacturing. Springer, Cham.

Patnaik, P. K., Swain, P.T. R., & Purohit, A. (2019). Selection of composite materials for structural applications through MCDM approach. *Materials Today: Proceedings*, 18, 3454–3461.

Prasad, K., & Chakraborty, S. (2014). A decision-making for non-traditional machining processes selection. *Decision Science Letters*, 3, 467–478.

Prasad, K., & Chakraborty, S. (2018). A decision guidance frame work for non-traditional machining process selection. *Ain Shams Engineering Journal*, 9, 203–214.

Rai, D., Jha, G. K., Chatterjee, P., & Chakraborty, S. (2013). Material selection in manufacturing environment using compromise ranking and regret theory-based compromise ranking methods: A comparative study. *Universal Journal of Materials Science*, 1, 69–77.

Raigar, J., Sharma, V., Srivastava, S., Chand, R., & Singh, J. (2020). A decision support system for the selection of an additive manufacturing process using a new hybrid MCDM technique. *Sadhana*, 45, 1–14.

Rao, V. R., & Lakshmi, J. (2021). R-method: A simple ranking method for multi-attribute decision-making in the industrial environment. *Journal of project management*, 6, 223–230.

Roth, R., Field, F., & Clark, J. (1994). Materials selection and multi-attribute utility analysis. *Journal Computer-aided Materials Design*, 1, 325–342.

Roy, B. (1968). Classement et choix en présence de points de vue multiples. *RAIRO-Operations Research-Recherche Opérationnelle*, 2, 57–75.

Saaty, T. L. (1980). *The Analytic Hierarchy Process: Planning, Priority Setting, Resources Allocation*. New York: McGraw.

Sadhu, A., & Chakraborty, S. (2011). Non-traditional machining processes selection using data envelopment analysis (DEA). *Expert System with Applications*, 38, 8770–8781.

Sahu, A. K., & Sahu, N. K. (2015). Benchmarking CNC machine tool using hybrid- fuzzy methodology: A multi-indices decision making (MCDM) approach. *International Journal of Fuzzy System Applications*, 4, 28–46.

Sen, B., Bhattacharjee, P., & Mandal, U. (2016). A comparative study of some prominent multi criteria decision making methods for connecting rod material selection. *Perspectives in Science*, 8, 547–549.

Shukla, A., Agarwal, P., Rana, R. S., & Purohit, R. (2017). Application of TOPSIS algorithm on various manufacturing processes: A review. *Materials Today: Proceedings*, 4, 5320–5329.

Singh, R. P., Tyagi, M., & Kataria, R. (2019). Selection of the optimum hole quality conditions in manufacturing environment using MCDM approach: A case study. *Operations Management and Systems Engineering*. In: Sachdeva, A. et al. (eds.), *Operations Management and Systems Engineering*, 133–152, Lecture Notes on Multidisciplinary Industrial Engineering. Springer, Cham.

Taha, Z., & Rostam S. (2011). A fuzzy AHP-ANN-based decision support system for machine tool selection in a flexible manufacturing cell. *The International Journal of Advanced Manufacturing Technology*, 57, 719–733.

Tian, G., Zhang, H., Feng, Y., Wang, D., Peng, Y., & Jia, H. (2018). Green decoration materials selection under interior environment characteristics: A grey-correlation based hybrid MCDM method. *Renewable and Sustainable Energy Reviews*, 81, 682–692.

Yazdani, M., & Payam, A. F. (2015). A comparative study on material selection of microelec-tromechanical systems electrostatic actuators using Ashby, VIKOR and TOPSIS. *Materials and Design*, 65, 328–334.

Yazdani, M., Zavadskas, E. K., Ignatius, J., & Yazdani, M. D. A. (2016). Sensitivity analysis in MADM methods: Application of material selection. *Inzinerine Ekonomika-Engineering Economics*, 27, 382–391.

Yurdakul, M., & Tansel, Y. (2009). Analysis of the benefit generated by using fuzzy numbers in a TOPSIS model developed for machine tool selection problems. *Journal of Materials Processing Technology*, 209, 310–317.

Yuvaraj, N., & Kumar, M. P. (2015). Multi response optimization of abrasive water jet cutting process parameters using TOPSIS approach. *Material and Manufacturing Processes*, 30, 882–889.

Zaman, K.U, Rivette, M., Siadat, A., & Mousavi, S. V. (2018). Integrated product- process design: Material and manufacturing process selection for additive manufacturing using multi-criteria decision making. *Robotics and Computer-Integrated Manufacturing*, 51, 169–180.

Zavadskas, E. K., Kaklauskas, A., & Sarka, V. (1994). The new method of multicriteria complex proportional assessment of projects, *Technological and Economic Development of Economy*, 1, 131–139.

Zavadskas, E. K., & Turskis, Z. (2010). A new additive ratio assessment (ARAS) method in multicriteria decision-making. *Technological and Economic Development of Economy*, 16, 159–172.

Zavadskas, E. K., Turskis, Z., Antucheviciene, J., & Zakarevicius, A. (2012). Optimization of weighted aggregated sum product assessment. *Elektronika ir elektrotechnika*, 122, 3–6.

Zavadskas, E. K., Govindan, K., Antucheviciene, J., & Turskis, Z. (2016a). Hybrid multiple criteria decision-making methods: A review of applications for sustainability issues. *Economic Research-Ekonomska Istraživanja*, 29, 857–887.

Zavadskas, E. K., Ghorabaee, M. K., Turskis, Z., & Antucheviciene, J. (2016b). A new combinative distance-based assessment (CODAS) method for multi-criteria decision making. *Economic Computation and Economic Cybernetics Studies and Research*, 50, 25–44.

Zhao, R., Su, H., Chen, X., Yu, Y., Wang, B., Zhang, N., & Rosen, M. A. (2016). Commercially available materials selection in sustainable design: An integrated multi-attribute decision making approach. *Sustainability*, 8, 1–15.

Chapter 4

To implement Six Sigma to minimize defects in the manufacturing of draft gear of railway wagon

Chanchal Singh, Deepak Agarwal, Anurag Singh, and Mayank Agarwal

CONTENTS

4.1   INTRODUCTION

In the modern era, Six Sigma's robust quality management tool is required to make a strong business strategy. Sigma is also a business management tool that improves the quality of production processes by minimizing and eventually removing errors and variations (Kumar et al., 2007). Various engineering and management tools have been developed to respond to customer requirements, like **Just in Time** (JIT) manufacturing, Enterprise Resource Planning, ISO 9000, **Total Quality Management** (TQM), and lean manufacturing (Furgał & Cygan, 2009).

The study is focused on the applicability of Six Sigma (DMAIC approach) to the manufacturing of side buffers and deals with the quality issue of components. It can be understood using its methodology and Six Sigma levels with DPMO

DOI: 10.1201/9781003293576-4

Table 4.1 Relation between DPMO and Process Yield

Sigma level	DPMO	Process yield
1	500,000	50%
2	308,537	65%
3	66,807	93%
4	6,210	99.4%
5	233	99.976%
6	3.4	99.9997%

(Defective Parts-Per Million Opportunity) and process yield (Sachin & Dileeplal, 2017; Girmanová et al., 2017). Table 4.1 shows how DPMO relates to process yield.

Six Sigma depends on a statistical enhancement tool designed to enhance business processes by reducing process variation. Quality has been an essential discriminating factor for many centuries now as the production industry was the first to take a hard, scientific tool at Quality (Kumar et al., 2009). Like defects in other manufacturing processes, defects from sand casting may be caused by many different factors; so, numerous experiments are usually required while implementing the Six Sigma methodology. The Taguchi method requires fewer experimental trials compared to the complete factorial design of the experiment model. This has made the Taguchi method a more economical methodology to integrate with Six Sigma to optimize product quality and increase benefits. Using the Taguchi method to optimize surface roughness in squeeze castings, Kim, Shin, Lee, and Moon caused the Taguchi method to optimize the die-casting conditions of alloy Az91D components (Chen & Brahma, 2014). According to 'Genichi Taguchi,' cost is more important than Quality, but Quality is the best way to control the cost. **Manufacturing industries** measure Quality prior in terms of measurements, and nowadays measurement is done using statistical methods. Seeking customer expectations, Six Sigma understands the elements of waste. Six Sigma (Yadav et al., 2021) (DMAIC methodology), in Figure 4.1, skills are used to achieve sustainable quality improvement through process improvement. Most casting defects arise due to the uncontrolled process parameter; hence, the concern is to produce the defect-free parts (zero defects) (Ganganallimath et al., 2019).

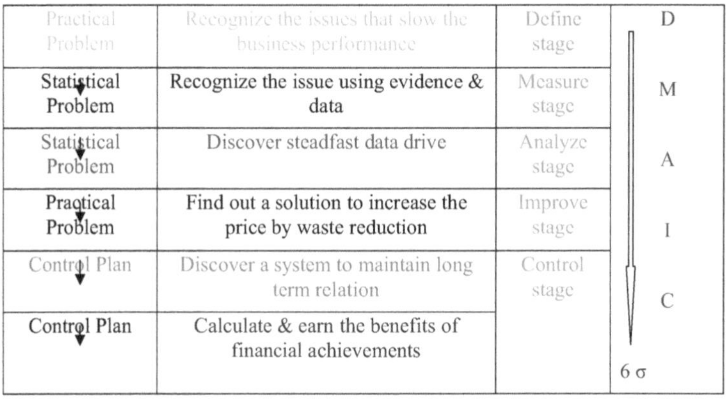

Practical Problem	Recognize the issues that slow the business performance	Define stage	D
Statistical Problem	Recognize the issue using evidence & data	Measure stage	M
Statistical Problem	Discover steadfast data drive	Analyze stage	A
Practical Problem	Find out a solution to increase the price by waste reduction	Improve stage	I
Control Plan	Discover a system to maintain long term relation	Control stage	C
Control Plan	Calculate & earn the benefits of financial achievements		6 σ

Figure 4.1 Asystematic approach of DMAIC (Qayyum et al., 2021).

4.2   BUSINESS PROCESS MAPPING (SIPOC DIAGRAMS)

4.2.1   Purpose

The acronym SIPOC stands for Suppliers, Inputs, Process, Outputs, and Customers. Process mapping is a vital business process skill in operation enhancement that utilizes the hieroglyph in the flow chart to get a visual sketch of business activity from start to end. This hieroglyph traces each step in a process and finds out what is being done and by whom, where, when, how, and even makes it possible to uncover why (Fitriana et al., 1978).

4.2.2   Definitions

Supplier: A supplier is a body that distributes a product or service from one end to the other end. The role of a supplier (Kannan, 2002) in a business is to dispense high-quality products from production at a reasonable price to a distributor or retailer for resale. A supplier in an industry is a body that acts as an intermediary between the manufacturer and the retailer, ensuring that conveying is forthcoming and stock is of sufficient quality. It may be an insider or outsider supplier.

   Input: Material, resources, and data required to execute the process

   Process: A chain of actions, motions, or processes leading to the production *process* or a collection of operations that take one or more kinds of input and develop output worth to the consumer

   Output: The tangible products or services that result from the operation

   Customer: Whoever receives the outcome of the process. It may be either internal or external.

4.3   DEFINE PHASE

4.3.1   SIPOC logic

A SIPOC (Ketabforoush & Abdul Aziz, 2021) diagram (Figure 4.2) is a visual tool for documenting a business process from beginning to end before implementation. SIPOC (pronounced sigh-pock) diagrams are also referred to as high-level process maps because they do not contain much detail. Before the process can be investigated, we shall prepare a roadmap in the form of a project charter as shown in Table 4.2 below (Gavriluţă, 2017; Parkash & Kaushik, 2011):

   A SIPOC diagram can be given as shown in Figure 4.2.

4.4   MEASURE PHASE

Pareto diagram: This is the data collection phase and data of casted draft gear housing from April 2019 to December 2019 in the greensand casting production line to investigate the problem this particular item faces, as shown in Table 4.3. The following table

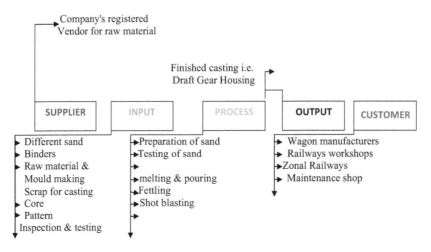

Figure 4.2 SIPOC logic layout.

Table 4.2 Sequence of a project charter

Project charter
Project title to reduce the rejection rate of the draft gear housing
Project objective to reduce present defect %
Critical to Quality (CTQ) rejection % is high due to casting defects.
Project scope: the green sand casting process
Expected benefit: quality product and defect-free product Customer satisfaction: cost-saving due to reduced defect %
Schedule:
Define - 1 week
Measure - 2 week
Analyze - 3 week
Improve - 3 week
Control - 3 week

Table 4.3 Production–rejection statement for draft gear housing

Name of month	Quantity produced	Quantity defective	Defective percentage
April19	2,600	105	4.03
May19	2,630	95	3.61
June19	2,650	107	4.03
July19	2,600	98	3.76
August19	2,620	105	4.0
September 19	2,680	95	3.54
October19	2,700	115	4.25
November 19	2,720	108	3.97
December 19	2,750	120	4.36
Total	**23,950**	**948**	**3.95**

shows the draft gear housing's total production–rejection statement (Ouyang & Wu, 1994; Wang et al., 2018).

With the help of Table 4.4, a Pareto chart has been drawn. This chart reveals the major contributor to defect count and type of defects as shown in Figure 4.3.

In the Pareto chart, the vertical axis of the left-hand side represents the total defective count, and the right-hand side represents the cumulative frequency of defect %. In contrast, the horizontal axis of the Pareto chart represents the response variable, i.e., type of defect. It can be seen with the help of the above Pareto diagram that scab contributes 43.77% of the fault, whereas sand inclusion contributes 34.28%. Scab and sand inclusion cumulatively is the prime cause for 78.05% of defective casting from the Pareto chart. Among the above-stated defects, **scab** causes the highest % rejection and

Table 4.4 Pareto analysis

S. no.	Type of defect for draft gear housing	Quantity defective	Cumulative sum	Cumulative frequency %
I	Scab	415	415	43.77
2	Sand inclusion	325	740	78.05
3	Mismatch	68	808	85.23
4	Misrun	52	860	90.71
5	Distortion	49	909	95.88
6	Others	39	948	100
Total		**948**		

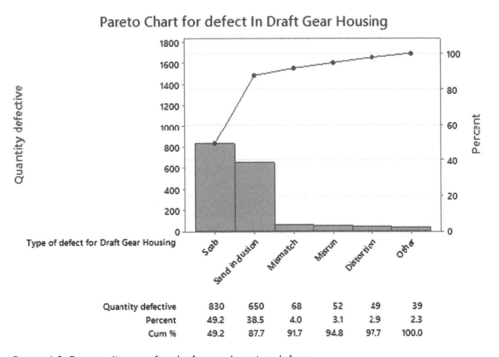

Figure 4.3 Pareto diagram for draft gear housing defect.

creates a barrier to achieving the production target. Now, we have to apply the statistical quality control tools to calculate the current sigma level and process yield of the firm using the DPMO formula and draw the control charts to understand the process capability of the firm.

4.4.1 DPMO

Using Table 4.2, to calculate DPMO, there are three basic pieces of information to be needed as given below:

i.   The number of units produced = 23,950
ii. The number of defect opportunities = 06
iii. The number of defects = 948

Now,

$$\text{DPMO} = \left(\frac{\text{No. of defective units}}{\text{No. of opportunities for defect} \times \text{No. of units produced}} \right) \times 10^6$$

$$= \frac{948}{06 \times 23,950} \times 10^6$$

$$= 6,597$$

Now, we have to calculate the Process Yield of the Casting Process

$$\text{Yield} = \left(\frac{\text{No. of opportunities for defect} \times \text{No. of units produced} - \text{No. of defects}}{\text{No. of opportunities for defect} \times \text{No. of units produced}} \right) \times 10^2$$

$$= \frac{06 \times 23,950 - 948}{06 \times 23,950} \times 10^2$$

$$= 99.34\%$$

To calculate, sigma level for the above DPMO & process yield using a standard table known as process sigma Table 4.5 (correlation table) & can be given as

Based on the above DPMO (Coskun et al., 2016; Setijono, 2009) and Process Yield (known as the sigma level calculator), the sigma level of the process is 4.0. After precise calculation, keeping each thing in mind, the baseline status can be tabulated as Table 4.6:

Table 4.5 Process sigma table

Sigma level	Defect rate	Yield
2 σ	308,770 DPMO	69.10000%
3 σ	66,811DPMO	93.33000%
4 σ	6,210DPMO	99.38000%
5 σ	233DPMO	99.97700%
6 σ	3.4DPMO	99.9997%

Table 4.6 Baseline status or Current performance of firm for draft gear housing

Part name	Average defect %	Process yield in %	DPMO	Sigma level
Draft gear housing	3.95	99.34	6,597	4.0

4.4.2   Control chart

Here we are going to draw the control chart for attributes (No. of defective pieces). Attribute charts are a set of control charts designed explicitly for Attribute data (i.e., counts data). Attribute charts monitor the process location and variation over time in a single chart. The brief classification of control charts for variables and attributes can be given as shown in Figure 4.4.

So, the keen interest is to draw the 'np' chart (Wu & Wang, 2007; Chong et al., 2014). The control chart for the current process can be drawn in the following manner in Table 4.7:

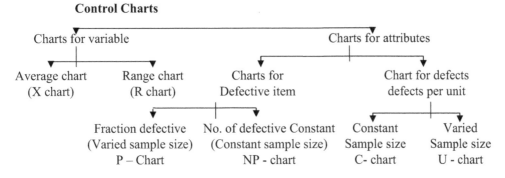

Figure 4.4 Classification of control charts.

Table 4.7 Control chart analysis for no. of defective pieces

S. no.	Month	Sample size	Defective pieces
1	April19	1,850	80
2	May19	1,850	78
3	June19	1,850	73
4	July19	1,850	79
5	August19	1,850	98
6	September 19	1,850	71
7	October19	1,850	94
8	November 19	1,850	76
9	December 19	1,850	99
Total		**16,650**	**748**

The calculation is done as follows:

i.   Center Line of the process (**CL**) $= np$

where $\text{Proportion defective} = \dfrac{\text{Total no. of defective pieces}}{\text{Total no. of Production (Sample Size)}}$

And $n =$ Sample Size $= 1,850$

Now $p = \dfrac{748}{16,650} = 0.04492$

Then **CL** $= 0.04492 \times 1,850 = 83.11$

ii. Upper Control Limit of the process (**UCL**) $= np + 3\sqrt{npq}$
     where $q = 1 - p$,
     $q = 1 - 0.04492 = 0.95508$
     Now UCL $= 1,850 \times 0.04492 + 3\sqrt{1,850 \times 0.04492 \times 0.95508}$
     UCL = 109.84

iii. Lower Control Limit(**LCL**) $= np - 3\sqrt{npq}$
     LCL $= 1,850 \times 0.04492 - 3\sqrt{1,850 \times 0.04492 \times 0.95508}$
     LCL = 56.38

Based on the above calculation, we can draw the control chart of the current manufacturing process of draft gear housing, as shown in Figure 4.5.

It is clear from the above-obtained control chart that all data points fall between UCL and LCL; so, we can say that process is in control.

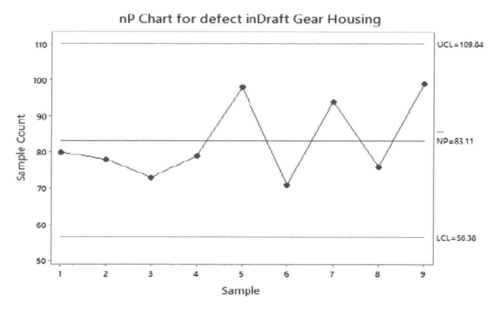

Figure 4.5 Control chart for production of draft gear housing.

4.5   ANALYZE PHASE

4.5.1   Root cause analysis

At this stage, a practical problem will get converted into a statistical problem and analyzed and then, planning and conducting brainstorming sessions by improvement team members (including quality assurance engineers) in the casting unit. Based on the discussions upon the probable causes (including major and minor causes) as to why the scab problem occurred, five significant factors may cause the defect:

i.   Man
ii.   Machine
iii.   Material
iv.   Method
v.   Environment

4.5.2   Defect name: Scab

A scab is a condition where an excess metal layer is found on the casting surface and can be readily removed by scraping or peeling. Under it is located a layer of sand on the casting surface. The primary cause for scabs is the sand mixture with either too much clay or moisture content to permit the sand to expand properly when the molten metal comes in contact with it. By way of correction, sand properties should be checked, and modifications made in the amount of bentonite and or water are added.

Another cause of scabbing is mold rammed too hard. In this case, the mold is rammed so hard that there is no room for the sand to expand when exposed to the molten metal heat. At this point, the mold face buckles, and the molten metal is allowed to penetrate behind the surface layer of the mold face. The correction for this condition is to lower the mold hardness by lighter ramming. The cause and effect diagram for scab is shown in Figure 4.6.

In the analysis phase, while finding the root cause of defect 'scab', many factors seem responsible primarily, but among those, vital are few and trivial are many. Hence, the matter was discussed with experienced foundrymen and quality control engineers to establish the root causes of the scabbing defect.

A systematic approach has been adopted. The standard operating range of various factors like grain size, clay content %, moisture content %, hardness, and temperature that involve in sand casting is shown in Table 4.8.

As per the standard sampling process, samples were collected at regular intervals of two days and tested for sand grain size, moisture content, and clay content in the laboratory. The temperature of molten metal is also checked. The hardness of mold samples at different locations was also checked to ascertain whether ramming was proper and uniform all over. Results obtained are tabulated in Table 4.9.

One more observation is needed that leads the attention toward the pouring speed. And need to check whether it is in control, which results in the erosion of sand near the gating resulting in the scabbing defect.

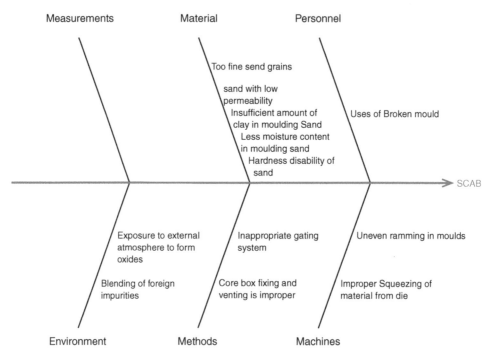

Figure 4.6 Cause and effect diagram for draft gear housing.

Table 4.8 Operating range of factors

S. no.	Grain size (**American Foundry Society**, AFS)	Clay content %	Moisture content%	Hardness **Brinell Hardness Number** (BHN)	Temperature°C
I	40–50	9–12	3.5–5	75–90	1,610–1,625

Table 4.9 Analysis of various factors causing defect 'scab'

Sample	Grain size — System sand (AFS)	Clay content % — System sand	Moisture content %	Hardness BHN	The temperature of molten metal in °C	Defect identified (Y/N)
1	46	19	4.62	80	1,615	Y
2	45	16	4.76	79	1,620	Y
3	47	15	4.59	81	1,614	Y
4	48	12	4.68	82	1,616	N
5	50	11	4.22	80	1,621	N

Table 4.10 Pouring time measurement

S.no.	Pouring time (seconds)	Deviation identified (Y/N)
1	43.02	N
2	43.09	N
3	43.02	N
4	43.05	N
5	43.04	N

To verify the pouring time in particular five molds was measured with a digital stopwatch. The empirical calculation (calculation done by foundrymen and based on their experience) says 100 kg mold takes 10 seconds to fill the mold cavity. Therefore, the weight of draft gear housing is approximately 143 kg. The pouring time using a stopwatch for different molds and their consequence is tabulated as shown in Table 4.10.

The pouring rate was correct as per usual practice from the above observation. It was observed from Table 4.9 that all these samples have different values of grain size, clay content, moisture content, green hardness number, and temperature; all matters meet the standard operating range; after that, corresponding results had displayed. The conclusion can be drawn from the above observation table except for clay content %, and all values meet the standard operating range criteria. So, it can be deduced that the clay content % was the major contributing factor to defect occurrence.

4.6 IMPROVE PHASE

This step aims to identify, test, and develop the optimal solution and implement the solution to check the confirmation in pilot production. In the analysis phase, the brainstorming session was planned and conducted to identify the root cause of the defect scab. To improve, phase focus is on developing the rid of the root cause of variation, testing, and standardizing the solution. Making effective discussion with the production department, quality department, and supervisors in the analysis phase, the probable cause of defect 'scab' is not to meet the clay content % with the standard in molding sand/system sand due to which defect appears at the in-gate of the mold.

After conducting brainstorming sessions with the quality control department and expert foundrymen, the outcome had arrived that after every molding operation, there is a need to perform a test of clay content % in the sand and try to control the value in the range of standard value, i.e., 9–12% to avoid the defect scab. Inline rejection is due to defect scab during the period 15 February 2020 to 15 March 2020. To see the trend of defect appearance in the production line of the component draft gear housing, the date-wise data of defect scab can be tabulated as shown in Table 4.11.

Table 4.11 shows the inline rejection due to scab per day during the period of 15 February 2020 to 15 March 2020, which indicates that there is an abrupt decrease in the inline rejection due to scab. The significant change is due to the use of the appropriate amount of clay content % in system sand. This table concludes that the use of an appropriate amount of clay content % will give a better result which is a goal of the DMAIC methodology.

Table 4.11 Inline rejection table date-wise for draft gear housing

February 2020	Defective piece	March 2020	Defective piece
15	2	1	1
16	1	2	1
17	1	3	NIL
18	1	4	1
19	1	5	NIL
20	1	6	NIL
21	1	7	1
22	NIL	8	NIL
23	1	9	NIL
24	1	10	1
25	NIL	11	1
26	NIL	12	1
27	1	13	NIL
28	1	14	NIL
29	NIL	15	NIL

4.7   CONTROL PHASE

The major defects are investigated and reduced up to a certain extent. The real challenge is to make the consistency of processes and products. For this, the following are the necessary actions in the view of **control plan** that has been taken by the organization:

i.   Use an adequate amount of clay and moisture in the sand.
ii. Arrange training and counseling sessions for concerned people involved in the molding line.
iii. Regular monitoring of sand in a laboratory is required.
iv. Maintaining the pouring speed of molten metal is required.

In addition, to the control phase, the statistical quality control of any manufacturing process is necessary, for which the data is collected after the improvement phase for the month of 15 February to 15 March 2020, as shown in Table 4.11.

Now, we have to calculate the current sigma level and process yield of the firm using the DPMO formula. For this:

i.   The number of units produced = 2,170
ii. The number of defect opportunities = 06
iii. The number of defects = 22

Now,

$$\text{DPMO} = \left(\frac{\text{No. of defective units}}{\text{No. of opportunities for defect} \times \text{No. of units produced}} \right) \times 10^6$$

$$= \frac{22}{06 \times 2,170} \times 10^6$$

$$= 1,690$$

Now, we have to calculate the Process Yield of the Casting Process

$$Yield = \left(\frac{\text{No. of opportunities for defect} \times \text{No. of units produced} - \text{No. of defects}}{\text{No. of opportunities for defect} \times \text{No. of units produced}} \right) \times 10^2$$

$$= \frac{06 \times 2,170 - 22}{06 \times 2,170} \times 10^2$$

$$= 99.83\%$$

Based on the above DPMO and Process Yield (known as the sigma level calculator), the sigma level of the process is 4.4. After precise calculation, keeping each strategy in mind, the baseline status can be tabulated as shown in Table 4.12.

Now, it is necessary to check the process variation of the 5-week production run (15 February to 15 March 2020) of the casting unit, using statistical quality control, i.e., 'np' chart for the no. of defective pieces. The procedure is as follows. All related data can be tabulated as shown in Table 4.13.

The calculation is done as follows:

i. Center Line of the process $(CL) = np$

where Proportion defective $= \dfrac{\text{Total no. of defective pieces}}{\text{Total no. of Production(Sample Size)}}$

And n = Sample Size = 350

Now, $p = \dfrac{9}{1,750} = 0.00514$

Then, $CL = 0.00514 \times 350 = 1.8$

ii. Upper Control Limit of the process$(UCL) = np + 3\sqrt{npq}$

where $q = 1 - p$,

$q = 1 - 0.00514 = 0.99486$

Now, $UCL = 350 \times 0.00514 + 3\sqrt{350 \times 0.00514 \times 0.99486}$

$UCL = 5.815$

Table 4.12 Current production–rejection statement

Name of month	Quantity produced	Quantity defective	Quantity rejected	Defective piece %
February–March 2020	2,170	22	3	1.01

Table 4.13 Control chart analysis after improvement

S. no.	Name of month (15 February to 15 March 2020)	Sample size	Defective pieces
1	Week 1	350	3
2	Week 2	350	2
3	Week 3	350	1
4	Week 4	350	1
5	Week 5	350	2
Total		1,750	9

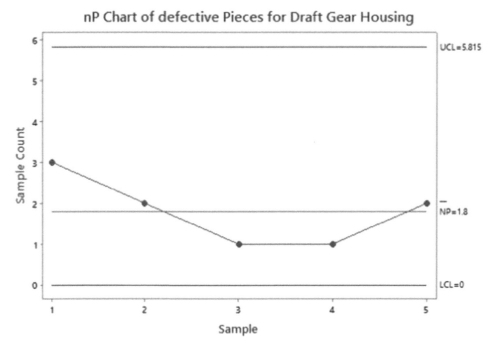

Figure 4.7 Control chart for draft gear housing after improvement.

iii. Lower Control Limit(**LCL**) $= np - 3\sqrt{npq}$

   LCL $= 350 \times 0.00514 - 3\sqrt{350 \times 0.00514 \times 0.99486}$
   LCL $= -2.2144$
   LCL = 0 (If LCL is negative, then it will be taken as zero.)

Based on the above calculation, we can draw the control chart of the current manufacturing process of draft gear housing, as shown in Figure 4.7.

   It is clear from the above control chart that all data points fall between UCL and LCL; so, we can say that process is in control.

4.8   RESULT AND DISCUSSION

In the present study, attention has been put on the defect investigation (a reason for the rejection of casted components) and their removal in draft gear housing. Six Sigma approach through DMAIC methodology is implemented with its various quality tools to analyze the data and implement the outcome (statistical analysis into graphical analysis). Six Sigma quality tools not only eliminate the defects in the casting process but also affect the process of knowledge creation (data collection and interpretation) and application positively. Process knowledge creation is a significant contributing aspect toward manufacturing process improvement.

By DMAIC approach, we found the percentage contribution of each scrap in our production and found the critical parametric defects (scab - 44%, sand inclusion - 34%, etc.). By this method, we found the root cause analysis of our product defect and used a systematic approach to eliminate these defects by a percentage of 75%. So, by this, we reduce the defective percentage in weekly samples by 60%.

4.9 CONCLUSION

In the globalization period, the casting industry is facing a tough challenge to produce a good quality product at low cost and to deliver in the stipulated leadtime to the customer. Six Sigma program is a powerful approach and, with its DMAIC methodology, creates an opportunity to handle this kind of quality problem situation. It identified the root cause of defects and eliminated them to improve customer satisfaction. There are many techniques (TQM, value stream mapping, line balancing, etc.)to solve this type of problem, but they are not capable to analyze systematically and lack data collection. The present study undertakes the Six Sigma approach through DMAIC methodology. It is a statistical graphical analysis tool. It attempts to reduce the 'scabbing' deficiency and reduce the rejection rate of casting product 'draft gear housing' in the foundry industry. Simultaneously, there is a chance for improvement regarding the elimination of other probable defects also.

REFERENCES

Chen, J. C., & Brahma, A. R. B. (2014). Taguchi-based six sigma defect reduction of green sand casting process: An industrial case study. *Journal of Enterprise Transformation, 4*(2), 172–188. https://doi.org/10.1080/19488289.2013.860415.

Chong, Z. L., Khoo, M. B. C., & Castagliola, P. (2014). Synthetic double sampling np control chart for attributes. *Computers and Industrial Engineering, 75*(1), 157–169. https://doi.org/10.1016/j.cie.2014.06.016.

Coskun, A., Oosterhuis, W. P., Serteser, M., & Unsal, I. (2016). Sigma metric or defects per million opportunities (DPMO): The performance of clinical laboratories should be evaluated by the Sigma metrics at decimal level with DPMOs. *Clinical Chemistry and Laboratory Medicine, 54*(8), e217–e219. https://doi.org/10.1515/cclm-2015-1219.

Fitriana, R., Saragih, J., & Sarasaty, S. (1978). Implementation six sigma and data mining to improve die casting production process at PT. AB. *Proceeding 7th International Seminar on Industrial Engineering and Management*, Jakarta, pp. 64–71.

Furgał, G., & Cygan, R. (2009). Quality problems root cause identification and variability reduction in casting processes. *Archives of Foundry Engineering, 9*(1), 13–16.

Ganganallimath, M. M., Patil, S. D., Gijo, E. V., Math, R. B., & Hiremath, V. (2019). Application of Taguchi-based Six Sigma method to reduce defects in green sand casting process: A case study. *International Journal of Business and Systems Research, 13*(2), 226–246. https://doi.org/10.1504/IJBSR.2019.098666.

Gavriluță, A. (2017). Analysis of a production system with the help of lean manufacturing tools. *TEHNOMUS - New Technologies and Products in Machine Manufacturing Technologies, 1*, 88–93.

Girmanová, L., Šolc, M., Kliment, J., Divoková, A., & Mikloš, V. (2017). Application of six sigma using DMAIC methodology in the process of product quality control in metallurgical operation. *Acta Technologica Agriculturae, 20*(4), 104–109. https://doi.org/10.1515/ata-2017-0020.

Kannan, V. R. (2002). Supplier selection and assessment. *The Journal of Supply Chain Management,* 11–21.

Ketabforoush, M., & Abdul Aziz, N. (2021). The effect of taguchi-based six sigma method on variation reduction in a green construction material production process. *Iranian Journal of Science and Technology - Transactions of Civil Engineering, 45*(2), 879–889. https://doi.org/10.1007/s40996-021-00583-1.

Kumar, M., Antony, J., Antony, F. J., & Madu, C. N. (2007). Winning customer loyalty in an automotive company through Six Sigma: A case study. *Quality and Reliability Engineering International, 23*(7), 849–866. https://doi.org/10.1002/qre.840.

Kumar, M., Antony, J., & Cho, B. R. (2009). Project selection and its impact on the successful deployment of Six Sigma. *Business Process Management Journal, 15*(5), 669–686. https://doi.org/10.1108/14637150910987900.

Ouyang, L. Y., & Wu, S. J. (1994). Prediction intervals for an ordered observation from a pareto distribution. *IEEE Transactions on Reliability, 43*(2), 264–269. https://doi.org/10.1109/24.295005.

Parkash, S., & Kaushik, V. (2011). LogForum ISO 9001 QMS in sports goods manufacturing industry. *Scientific Journal of Logistics, 7*(4), 1–16.

Qayyum, S., Ullah, F., Al-Turjman, F., & Mojtahedi, M. (2021). Managing smart cities through six sigma DMADICV method: A review-based conceptual framework. *Sustainable Cities and Society, 72*(March), 103022. https://doi.org/10.1016/j.scs.2021.103022

Sachin, S., & Dileeplal, J. (2017). Six sigma methodology for improving manufacturing process in a foundry industry. *International Journal of Advanced Engineering Research and Science (IJAERS), 4*(5), 2456–1908. www.ijaers.com.

Setijono, D. (2009). The application of modified "Defect Per Million Opportunities" (DPMO) and sigma level to measure service effectiveness. *International Journal of Six Sigma and Competitive Advantage, 5*(2), 173–186. https://doi.org/10.1504/IJSSCA.2009.025168.

Wang, Z., Jia, H., Xu, T., &Xu, C. (2018). Manufacturing industrial structure and pollutant emission: An empirical study of China. *Journal of Cleaner Production,* 197, 462–471. https://doi.org/10.1016/j.jclepro.2018.06.092.

Wu, Z., & Wang, Q. (2007). An NP control chart using double inspections. *Journal of Applied Statistics, 34*(7), 843–855. https://doi.org/10.1080/02664760701523492.

Yadav, N., Shankar, R., & Singh, S. P. (2021). Cognitive aspects of lean six sigma. In *Quality and Quantity* (Issue April). Springer Netherlands. https://doi.org/10.1007/s11135-021-01141-7.

Chapter 5

Changeover time reduction through SMED approach

Case study of an Indian steel processing centre

Abhijit Das, Sumer Sunil Sharma, Sushovan Ghosh,
Tara Charan Bharti, and Sandeep Mondal

CONTENTS

5.1 INTRODUCTION

In the last few decades, the steel manufacturing industry has become ever so competitive that even a minor misjudgment can lead to a massive slip in the global race. Among all the metals, steel holds the historically dominant position for being the driver of industrialization. Thus, it will never be an overstatement to mention that steel is the backbone of any economy and has always been an industrial development booster. Today's era of globalization and worldwide connectivity has made understanding customers' changing demands and flexibility in business processes an utmost priority, meaning the manufacturing sector must make a swift shift to small batch size production. The sole aim of making the shift is to provide the customer with products or services, of the desired quality, in the required amount, and in due time. Though on paper, the idea of shifting focus to flexibility and small batches looks appealing, these come with many complications. This type of manufacturing would lead to a fundamental change in the enterprise planning approach and a significant increase in the setup frequency. Implementing lean manufacturing philosophy would enable the organization to systematically identify and eliminate waste through continuous and comprehensive improvement and ultimately serve the goal of quick setup for flexibility

DOI: 10.1201/9781003293576-5

and small-size batch production. SMED is one such lean manufacturing tool that aims to and enables bringing down setup/changeover time to under 10 minutes. Goubergen and Van Landeghem sighted three fundamental reasons why setup time reduction is vital for any manufacturing organization. These are also the key reasons why more and more industries are getting interested in the quick changeover, firstly for inventory reduction and gaining flexibility. Secondly, it could be so that organization is trying to identify and reduce the bottleneck capacity. Often, buying a new machine to meet the demand is expensive; so, the ultimate solution could be to increase the production rate by decreasing the setup time and operating the available machine at total capacity, provided the device is currently not fully utilized. Therefore, the bottleneck comes up to be a priority for reduction of setup time. The last reason for implementation is to minimize the overall cost as costs of the product are related to the effectiveness of equipment [1].

In the Indian steel processing scene, SMED has seen a limited application altogether due to several reasons and hence the project carried out is an effort to break the ice. SMED is mainly used to suppress the amount of time needed for the entire changeover, while other tools like why/why analysis, Poka-Yoke, Kaizen, etc., are used to maintain a standardized time and limit the variation in processes.

5.2 LITERATURE REVIEW

SMED was first developed by a Japanese Industrial engineer, Shigeo Shingo, in the 1950s. Shigeo Shingo conducted a production efficiency improvement study at the premises of Toyo Kogyo's Mazda plant in Hiroshima, Japan, where the concept of SMED first emerged; however, it was only in the seventies that it was widely acknowledged, given the fact that it was part of Toyota Production system [2]. Shingo ultimately became extraordinarily successful in reducing the changeover time for many companies. The aim of SMED is basically to bring down the setup and changeover time to under 10 minutes. Not all setup processes carried out throughout the industry can be performed in under 10 minutes, but this theory helped reduce the time drastically [3]. The study of the single-minute exchange of the die was initially developed by the study of the die change process, hence the name. Initially, SMED was developed primarily to counter the emerging need to increase smaller lot size production to satisfy flexible market demands. A quick change over the equipment can do wonders for production flexibility [4].

According to Shingo, the implementation of SMED can bring about plenty of positive gains for manufacturing organizations, directly or indirectly. Benefits in the form of reduction of error during the setup process, product quality improvement, and less time required for fine-tuning of machines and equipment can be known to be direct, while other types of rewards could be the rationalization of tools, reduction in inventory levels, and, most importantly, an increase in production flexibility [5].

Much of the SMED application can be traced back to the 1950s when Shingo at Toyota observed a few inefficient operations in the body-moulding process. The majority of the waste observed came from time spent changing dies of three large body-moulding presses and switching tools needed for different phases in the processes. He even noticed that the changeovers of tools on 500-tonne presses used to produce body panels took several hours to even days. Toyota also had to spend a massive amount of capital on land for the vehicles being built. Shingo again suggested that speeding up the changeover process could save Toyota a considerable amount of money. By the time it was the 1970s, the changeover time was reduced to just 3 minutes. The journey wasn't that smooth, and it encountered many hurdles in the form of complications and modifications. The difficulties came in changing the entire process to require less time for a changeover, modifying factory equipment and body parts of vehicles; Toyota also had to change the whole order of steps to build the car body mouldings. All these changes contributed significantly to implementing the "just-in-time" philosophy that Toyota had always wanted to execute to run smaller batches without affecting productivity. SMED plays a vital part in just-in-time manufacturing which reduces the time between the customer's demand/order and the ability of the organization to fulfilling it [2].

Yashuhiro Monden, in 1983, emphasized the adoption of parallel activities in production plants. He also backed up the idea of adopting setup process mechanization and the total elimination of adjustments. Any shop floor that involves setup work on both sides or on the front and backside of the machinery should apply two or more workers operating simultaneously. If only one particular operator is given the task for all the setup activities, then much of the time is wasted as the worker has to make frequent movements from one side to another or from back to forth. Monden describes the production as a network of operations and processes. Hence, transforming raw materials into the desired product is done by performing a series of operations. Four broad processes, namely processing, inspection, transportation, and delay, are carried out for the transformation of materials into finished goods, wherein delays can again be divided into two categories as process delay and lot delay. In process delay, the entire lot has to wait until the previous lot is processed and moves away, while in case of lot delay, another piece waits while one is being processed. Increasing lot size can result in quality reduction and material waste as it is challenging to pinpoint defects in a large lot compared to small ones. SMED is one such tool that can significantly reduce the setup time and reduce the need for larger lot sizes.

The application of lean is not limited to manufacturing industries; it has shown positive results in other ventures as well. Maalouf and Zaduminska [6] conducted the implementation of lean tools in the food processing industry and contrary to the historical literature and trends, the company has managed to reduce changeover time by 34%, adding to that the production capacity of the main production line which constitutes 84% of the total production of the concerned company has increased by 11%. As organizations like food processing industries possess unique characteristics in the form of heterogeneous raw materials, seasonality, short shelf-life, complex resources, etc., the implementation and impact of lean practices in these industries are limited. Furthermore, Maalouf and Zaduminska have made a distinction between discrete and process industries. In discrete industries, the products manufactured are individual units, for example, automobile, mobile handsets etc., which involve operations to be performed on individual items or groups of similar items. On the other hand, process industries produce bulk items, and individual separation is complex (e.g., steel, food products, and chemicals). Inflexible processes and high fixed capital are a few characteristics that process industries possess, which has hindered management from adopting lean practices. The concept of takt time is challenging to implement in process industries as takt time is already defined by machinery design in these types of industries [6].

In a study carried out by Monteiro et al. on a metalworking industry for improving the machining process, the prime objective was to increase productivity and eliminate waste. Several key processes were identified and mapped where SMED came up to be the major lean tool for achieving the improvements. The setup time was reduced by 40% on the vertical milling machine, while a 57% reduction was recorded on the horizontal milling machine. The study adopted a methodology known as Action-Research, which requires planning, acting, and observing the daily activities on the shop floor more carefully. This helps inculcate a thorough knowledge on the part of workers regarding their regular work and hence induces a will for improvement [7]. Another study was conducted on an electron beam machine in an automotive supplier company where implementation of SMED helped them reduce the setup time by more than 50%. The first step towards implementing the lean tool was forming a team that included personnel from departments, namely production, maintenance, process engineering, health and safety, and industrial and plant manager. This was followed by data collection, diagnosing the setup operation that had the most impact, and implementing core SMED principles [8].

Karam et al. carried out research on implementing SMED in a pharmaceutical company. SMED methodology, as a lean tool, was used at a bottle filling line in a Romanian pharmaceutical company to improve the changeover process. Implementing lean principles not only helped decrease the changeover time at the bottleneck process by up to 30% but also brought in added advantages in the form of process quality improvement, standardization of work, and not to mention several economic benefits. These positive results were seen just 12 months from the date of implementation. For the successful implementation of SMED, mapping the define-measure-analyze-improve-control (DMAIC) structure proved to be vital [9]. Vieira et al. [10] conducted a study on the application of SMED in an automotive component manufacturing company that produces air conditioning pipes. The main concern was to reduce the extremely high setup time of the deep drawing machine by about 20%. Deep drawing machine is crucial for the production of tubes, and thus quick setup of these machines is very much necessary. The setup time previously clocked at 47 minutes, and the machine availability was under 95%. The SMED implementation results came out to be quite satisfactory; not only did SMED help the management to standardize the setup operation, but the machine setup time was also drastically decreased by 38%, which was previously aimed at 20%. 53% of internal setup was reduced, and the overall equipment effectiveness availability increased by 7.7%

5.3 METHODOLOGY

This paper depicts a combined methodology to achieve the goals of reducing changeover time and its variability. The detailed methods of using SMED combined with other lean tools are described in Figure 5.1. SMED is the primary tool for reducing the setup and changeover time, while cause and effect analysis and Poka-Yoke are additional tools used to suppress the variability in changeover operation.

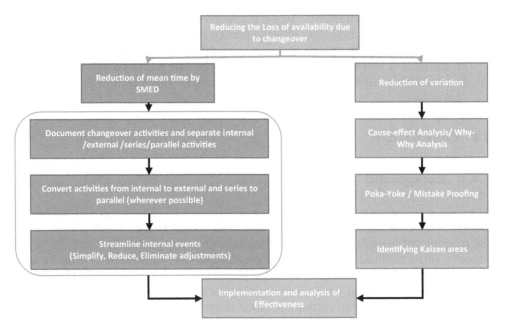

Figure 5.1 Methodology for reduction of changeover time and variability.

Note. The above methodology is exclusively designed by the authors involved in this project.

The processes involved in SMED can be categorized broadly into three steps, which are discussed as follows:

- The first step involves documenting and identifying activities in the entire changeover process and classifying them as internal/external or series/parallel activities.
- Next step would include converting internal activities to external ones. The main aim of SMED should lie in converting as many internal elements to external processes as possible. Each and every component should be thoroughly examined and checked if the process can be performed while the machine is still running; if yes, then convert it to an external process. An organization should also focus on and identify which of these modified steps have the biggest impact on time and cost; also, this step involves SMED to eliminate those steps which are non-value-adding.
- The final step deals with streamlining all the processes involved in the setup operation. Irrespective of internal and external setups, this final process emphasizes seeking systematic improvement; also, coming up with ideas to accomplish different tasks in an easier, faster, safer, and smoother manner is an utmost priority.

5.3.1 Principles of SMED

- **Changeover time:** The time taken for a change of a piece of equipment from producing the last good part of a production lot to the first good part of the next production lot.

- **Internal activities:** Internal activities are those activities that require the machine or process to be at a standstill. To put it another way, these are operations that can be performed either before or after the machine is shut/stopped.
- **External activities:** Activities that can be simultaneously done while the process or machine is still in running condition can be categorized under external activities. The machine need not be at a standstill or shut down to perform these operations.

5.4   CASE STUDY

5.4.1   Background

The study has been carried out in a steel processing centre of an integrated steel plant located in the eastern part of India. The steel coils are cut into narrower widths as per the requirement of customers in the slitting machine. The slitting line consists of the uncoiler, slitter machine, and recoiler. The hot-rolled steel coils received from the mill are mounted on the uncoiler, then uncoiled and fed to the slitter machine. The slitting machine's circular cutters perform the cutting action of sheets to the desired width. Finally, the output from the cutting operation is again fed to the recoiler, where sheets are rolled for easy storage and transportation.

Scrap choppers are equipment used in the slitter line to chop off scrap generated into smaller pieces for easy disposal and collection. In a regular slitting line, four choppers perform the task of chopping off the chips generated from the edge trim in the slitting of metal sheets. The knives of the chopper wear out regularly and thus require frequent changeovers. The major problem encountered from the side of the shop floor personnel was the tremendous amount of time required for the chopper changeover operation, sometimes as high as 9 hours for one particular changeover. Management initiated this project to reduce the time taken during the changeover activities and reduce the variability of time to increase the availability of the slitting machine. The baseline performance of the chopper change process was analysed by collecting the data for the previous 12 months. Data showed a significant amount of time for the changeover process with a mean time of 7.62 hours and a standard deviation of 1.23 hours. The target set under these circumstances was to reduce the mean and variability of chopper changeover time. (As shown in Figure 5.2)

5.4.2   Analysis

The project kicked off by mapping all the processes involved in the chopper changing process. The chopper geometry is shown in Figure 5.3, and all the activities involved in the changeover were thoroughly studied by analysing video recordings and conducting a method study. Table 5.1 shows all the processes involved in changing the choppers, from dismantling them to placing new ones on the shaft. Complete chopper change operation was studied in different shifts for a period of 1 month, manually employing a stopwatch by standard time study methods.

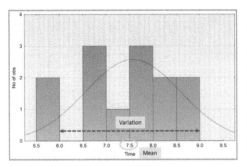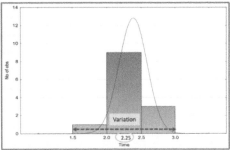

Figure 5.2 Baseline and target performance – analysis done by authors.

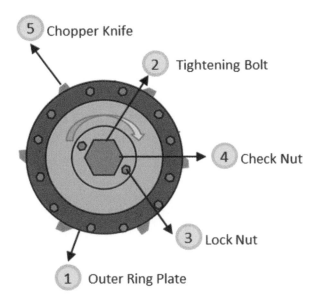

Figure 5.3 Chopper geometry – diagram prepared by the authors.

The next task was to figure out and eliminate all the non-value-adding activities. Separating out internal and external activities, series and parallel activities. This separation helped identify activities that can be performed when the machine is in operation or the activities that can be done in parallel with another activity. Other lean tools that helped to carry out the overall improvement process include 5S **(sort, set in order, shine, standardize and sustain)**, Poka-Yoke, and Kaizen (Figure 5.4).

Major issues detected after the analysis are:

- Activities like collecting tools, calling crane operators, and taking breaks in between are non-value-adding activities and should be eliminated.

Table 5.1 Time study of chopper changeover process

Activity	Moving of nuts	Type of operation	Min	Sec
Commencing the changeover			4	30
Collecting tools			6	12
Opening front ring plate	10	Automatic	13	37
Opening of the centre tightening bolt	1	Manual	32	19
Opening locking bolt of check nut	2	Automatic	9	15
Calling crane operator			3	11
Removing chopper with lock bolt	1	Manual	78	45
Taking away the chopper with crane		Automatic	11	57
Searching for fallen objects			6	51
Break			23	44
Attaching back ring plate	10	Automatic	21	13
Bringing chopper with crane and fixing on shaft		Automatic	11	56
Fixing the centre check nut and its locking bolt	2	Automatic	8	56
Fixing tightening bolt	1	Automatic	33	29
Attaching front ring plate	10	Automatic	11	28
Checking clearance between the cutting tools			28	11
Adding packing if required			10	11
Machine setting			7	9
Total			314	534

Note. The above table has been made by the authors involved after careful inspection of each and every process of changeover in the slitter chopper.

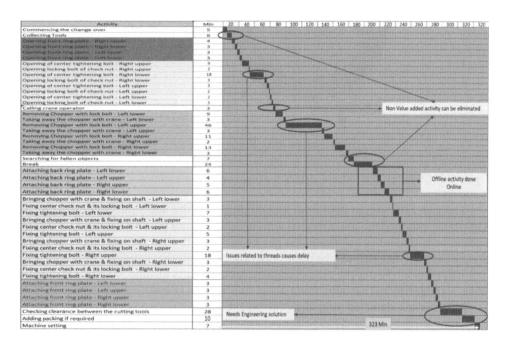

Figure 5.4 As-Is activity sequence of chopper changeover.

Note. The above is a Gantt chart prepared after analysing the as-is condition, made exclusively by the authors involved.

- Attaching the back ring plate for all the four choppers can be done offline which is being done online.
- Checking the clearance between the cutting tools requires an efficient engineering solution instead of adding packing and iterating the clearance for every combination of engagement of chopper knives.
- The shop floor lacks in overall supervision and added to that the layout of the shop floor needs to be revised.

5.4.3   Summary of recommendations

After carefully analysing all the data, charts, and figures, the project team came up with many ideas for improvement. Many sequential activities were shifted to parallel activities to save the overall time for the changeover. Internal and online activities were identified and changed to external activities, which can now be done before the commencement of the processes. One significant engineering change introduced is the use of a dial gauge to check the clearances of in-contact knives of the chopper for the cutting action (Table 5.2).

5.5   RESULTS AND DISCUSSION

The improvement ideas put forward by the project team were welcomed by the management and finally implemented. A revised SOP (Standard Operating Procedure) and activity sequence timeline were developed based on all the improvements suggested. Comparing the new chart directly with the baseline performance infer a reduction of time of 207 minutes. Earlier, the entire procedure for changeover took 323 minutes which after improvements is being reduced to only 116 minutes, as shown in Figure 5.5. The start of the project can be traced back to August 2019 when the average time for each changeover was measured at 7.66 hours before the commencement of SMED. The average monthly figures were as high as 9 hours per changeover for November 2019 and April 2020.

Finally, August 2020 was when the implementation of lean started and gradually showed positive results. The entire improvement process carried out on the shop floor had a massive impact on the bigger picture. First, the changeover time steadily decreased from 9.8 hours to mere 3.13 hours per changeover in July 2021. Increasing the availability of machines by 4.7% meant better throughput, which directly impacted the production rate.

All these factors mentioned above ultimately helped the organization make a revenue gain of up to 4.4 crore INR (0.5 million $).

5.6   CONCLUSION

The paper presents an optimal standard process for the changeover involved in the slitting line to reduce changeover time and its variability. The project's main goal was to reduce the 7.3% availability loss at the slitter line due to the changeover of choppers. The use of SMED has helped the plant counter this loss and eventually increase the

Table 5.2 Recommendations for improvement

Process critiques	Current time	Leads to	Solution approach	Enabler category	Additional resources required	Proposed time	Time saved
Parallel activities performed as sequential activities	114	Overall increase in changeover time	Two teams of two members to work on the left and right sets of choppers simultaneously	Work practice improvement	Gang of two workers, additional tools	56	58
Breaks and handover activity	24	Temporary stoppage of activities	Workers and supervisors to be provided	Work practice improvement		0	24
Bolts getting stuck in the cavity due to thread issues	82	Delay in screwing and unscrewing time	Setting proper matches after trials and colour coding them – Chopper Head, Check Nut, Tightening Bolt, Jacking Bolt De-threading and offline trials of the complete set, to check the thread issues. Proper lubrication before usage	Work practice improvement	Additional purchase of bolts and nuts	22	60
Attaching back ring plate to chopper head	22	Offline activity done online	Purchasing extra ring plates and attaching them to the chopper beforehand	Process improvement	Spare ring plates	0	22
Unnecessary time wasted in collecting tools	6	Delay to commence changeover	Creating trolley for tools and equipment, with 5S and ergonomics. Nuts and Bolts, Tools, Lubrication, and oil	System improvement	Inhouse designed and produced trolley	0	6
Time wasted in searching for nuts and bolts	5	Delay in second half of changeover	Creating trolley for tools and equipment, with 5S and ergonomics.	System improvement	Inhouse designed and produced trolley	0	5

(Continued)

Table 5.2 (Continued) Recommendations for improvement

Process critiques	Current time	Leads to	Solution approach	Enabler category	Additional resources required	Proposed time	Time saved
Searching for fallen objects	4	Non-value-added activity	Creating trolley for tools and equipment, with 5S and ergonomics.	System improvement	Inhouse designed and produced trolley	0	4
Low illumination level		Delay and variation	To avoid the shadow of the crane on the units, additional lights to be provided	System improvement	Additional lights		
Absence of overall supervision		Delay and variation	One person from each shift should be designated as "Chopper change in charge", Reverse Timer for tracking Pre-changeover checklist to be maintained by the in-charge to certify. Availability of well-maintained trolley, Availability of manpower resources, Illumination, and Crane operating conditions	System improvement Work practice improvement			
Designing proper layout		Delay and variation	Marking on floor separating two teams operating area, marking for trolley, Overhanging arrangement for air gun and screw gun	System improvement	Overhanging arrangement		
Checking clearance between the cutting tools and adding packing if required	38	Offline activity done online	Making sure the chopper knife's outer diameter is at the required level beforehand	Work practice improvement	Dial gauge	10	28

Note. The above table contains all the recommendations and solutions which are to be made to reduce the setup time. This table is exclusively made by the authors involved after careful analysis and brainstorming.

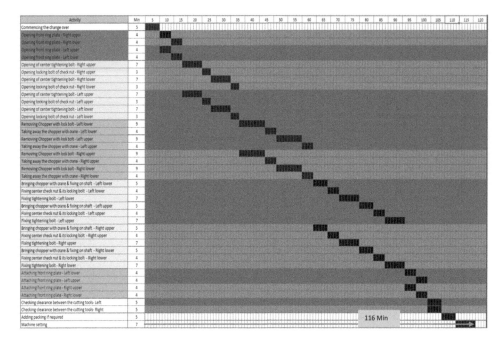

Figure 5.5 Revised activity-sequence-timeline (Gannt chart prepared by the authors).

availability by 4.7%, which resulted in a revenue gain of up to 4.4 crore INR. The application of lean was limited to only one machine within the plant. As the implementation has shown positive results, the next step would be to apply similar changes to all the similar slitting machines belonging to this manufacturing company. Horizontal deployment of lean methods to other machines in the same plant is also a possibility. This paper presents the use of SMED as a lean tool along with a few other lean tools such as 5S and Poka-Yoke. The future scope of research would include a broader focus on the use of SMED with many other lean tools, namely value stream mapping, plan do check and act (PDCA), total productive maintenance, etc.

REFERENCES

1. D. Van Goubergen, H. Van Landeghem (2002). Rules for integrating fast changeover capabilities into new equipment design. *Robotics and Computer-Integrated Manufacturing*, Vol. 18, No. 3–4, pp. 205–214.
2. A. C. Moreira, P. M. T. Garcez (2013, March). Implementation of the single minute exchange of die (SMED) methodology in small to medium-sized enterprises: A Portuguese case study. *International Journal of Management*, Vol.30, No. 1, p. 66, 60, 87.
3. Y. R. Mali, K. H. Inamdar (2012, May-Jun). Changeover time reduction using SMED technique of lean manufacturing. *International Journal of Engineering Research and Applications (IJERA)*, Vol. 2, No. 3, pp. 2441–2445. ISSN: 2248-9622.

4. B. Ulutas. (2011). An application of SMED methodology. *International Journal of Mechanical, Aerospace, Industrial, Mechatronic and Manufacturing Engineering*, Vol. 5, No. 7, pp. 1194–1197.

5. S. Shingo, A. P. Dillon (1985). *A Revolution in Manufacturing: The SMED System*, Productivity Press.

6. M. M. Maalouf, M. Zaduminska (2019, June). A case study of VSM and SMED in the food processing industry. *Management and Production Engineering Review*, Vol. 10, No. 2, pp. 60–68.

7. C. Monteiro, L. P. Ferreira, N. O. Fernandes, J. C. Sá, M. T. Ribeiro, F. J. G. Silva (2019). Improving the machining process of the metalworking industry using lean tool SMED. *8th Manufacturing Engineering Society International Conference*. 19–21 June, Madrid.

8. M. Martins, R. Godina, C. Pimentel, F. J. G. Silva, J. C. O. Marias (2018, June). A practical study of the application of SMED to Electron-beam machining in automotive industry. *28th International Conference on Flexible Automation and Intelligent Manufacturing (FAIM2018)*, June 11–14, 2018, Columbus, OH, USA.

9. A.-A. Karam, M. Liviu, V. Cristina, H. Radu (2017). The contribution of lean manufacturing tools to changeover time decrease in the pharmaceutical industry. A SMED project. *11th International Conference Interdisciplinary in Engineering, INTER-ENG 2017*, 5–6 October 2017, Tirgu-Mures, Romania.

10. A. M. Vieira, F. J. G. Silva, R. D. S. G. Campilho, L. P. Ferreira, J. C. Sá, T. Pereira (2021). SMED methodology applied to the deep drawing process in the automotive industry. *30th International Conference on Flexible Automation and Intelligent Manufacturing (FAIM2021)*, 15–18 June 2021, Athens, Greece.

Chapter 6

SWOT analysis – on maintenance frameworks for SMEs

Gouraw Beohar, Rajesh P. Mishra, and Anjaney Pandey

CONTENTS

6.1 INTRODUCTION

Maintenance framework implementation and optimization is always a tough task for researchers. SWOT helps them to highlight the existing strengths and weaknesses of a system, which guides the management to take suitable decisions regarding the maintenance system of the enterprise. This chapter establishes a SWOT analysis for various existing frameworks of maintenance to serve the important purpose of implementing maintenance in different SMEs. The different available frameworks are classified into four groups, namely A, B, C and D. Different groups are categorised depending on their methodologies: quantitative, qualitative, fundamental definitions, etc.

6.1.1 Requirement of a maintenance framework

Different SMEs have different maintenance framework requirements, which are discussed as follows:

- To contribute a generalized picture of the control of maintenance framework in SMEs
- To present an arrangement for the development of a pool of knowledge of various certified schemes and provide awareness of different maintenance frameworks
- To present a structure for evaluating the overall development of maintenance management system in an organization

DOI: 10.1201/9781003293576-6

- For evaluators and observers to explain different information, strategies, performance measurements and maintenance management
- To present the arrangement of competency outputs and solutions of the various components associated with maintenance management
- To generate an overall development system for enterprises

The prior literature on the maintenance framework was taken into consideration, and it was examined. SMEs are specifically mentioned in the literature collection, which also includes diverse maintenance strategies, as well as their choice, development, and classification on necessary topics. Identification of several frameworks that are available in published literature and their distinct parts. In order to examine the many types of maintenance frameworks accessible in published literature, it was initially necessary for all types of SMEs to gather data on various maintenance strategies—those beneficial for the construction of suitable maintenance frameworks.Our target is to increase the acquaintance with different framework approaches used in SMEs. To recognize the published literature's shortcomings for additional research in the area of maintenance like approaches, development and selection, we used some techniques like SWOT analysis on it.

6.1.2   Framework comparison (conceptual basis)

Paper considered 44 different maintenance frameworks over the last 21 years of time span which are available in previously published literature and compared all of them in the tabular format shown below in Table 6.1.It presents consolidated data of various maintenance frameworks simultaneously. Academicians consider different number of elements for various frameworks utilized in enterprises. Table 6.1 considers the comparison among all available data on framework type, pillar numbers, elements quantities, year, etc.
From Table 6.1, the following observations could be considered:

i.   Various researchers and academicians of different countries define the maintenance framework by a variable number of elements.
ii.   At the time of implementation of maintenance framework in SMEs, the range of elements is 3–14. Their quantity depends upon the type of SMEs, nature of work, turnover of enterprises, number of employees, etc.
iii.   From the table, it is observed that approximately 36% of the framework study is performed by Indian researchers. For the UK, the USA and Malaysia, the percentage is 10%, 7% and 9%, respectively.
iv.   Framework is utilized as an optimization tool which improves the overall efficiency of enterprises.
v.   Comparison table helps the researcher to get a clear idea about the existing types of frameworks in all types of SMEs.
vi.   Comparative analytical tabular data helps the author to implement SWOT analysis to find out gaps and barriers in the maintenance framework of SMEs.

Table 6.1 Comparison of frameworks

S. No.	Name of author	Year	Journal name	Country	Maintenance framework	No. of pillars	Name of element	Company
1	Levy, M., et al.	1999	Information Management	UK	ISS	3	Awareness, opportunity, positioning	Any SME
2	Labib, A. W.	1999	Management Decision	UK	Benchmarking, TPM	3	Data collection/select policy, data analysis/policy to machine, decision analysis	All organizations
3	Yusof, S. M. and Aspinwall, E.	2000	Total Quality Maintenance Magazine	UK	TQM implementation	4	Quality initiatives, general methodology, policy vision, coordinating body	Small scale industry
4	Kutucuoglu, K. Y., et al.	2001	International Journal of Operations & Production Management	UK	Performance measurement	7	Availability, reliability, utilization, schedule cost, breakdown cost / total maintenance cost, no. of delays, multi-skilled workforce	Manufacturing SMEs
5	Waeyenbergh, G. and Pintelon, L.	2002	International Journal Production Economics	Belgium	Reliability-centered maintenance	5	Startups / identification of objectives/ most important system / critical analysis, maintenance policy / PM policy, continuous improvement,	Production equipment
6	Venkatesh, J.	2005	Plant Maintenance Resource Center	India	Total Productive Maintenance (TPM)	8	Autonomous maintenance kaizen, PM, 5S, quality maintenance, training, TPM, safety and health	Nippondenso of Toyota
7	Chang, H. H.	2005	TQM and Business Excellence	Taiwan	Total Quality Management (TQM)	4	Total quality strategy, continuous improvement, performance factors, customer satisfaction / supplier-customer relation	All enterprises

(Continued)

Table 6.1 (Continued) Comparison of frameworks

S. No.	Name of author	Year	Journal name	Country	Maintenance framework	No. of pillars	Name of element	Company
8	Deros, B. M., et al.	2006	Benchmarking: An International Journal	Malaysia	Benchmarking, TQM	6	Top management vision, Critical success Factor / Human Resource Management / creativity, customer satisfaction, planning/analysis, benchmarking, reliability/ teamwork	Automotive company
9	Islam, M. A., et al.	2006	International Conference on Management of Innovation & Technology	New Zealand	Risk management	5	Identification of facility, identification of hazards, evaluation of risk, development of action, risk management	All SMEs
10	Kumar, M., et al.	2006	Production, Planning and Control	India	Six Sigma, TPM Lean	5	Define, measure, analyze, improve, control	Automobile accessories
11	Mishra, R. P., et al	2006	Journal of Advanced Manufacturing Systems	India	WCM (World Class Maintenance)	9	Productivity, cost, morale, reliability, quality, safety, effectiveness, flexibility, safety	Small scale industry
12	Baglee, D., et al.	2007	International Design Engineering Technical Conferences	UK	Maintenance management	5	Cost, function, time, overall equipment effectiveness, autonomous maintenance	All SMEs
13	Thomas, A. and Lewis, G.	2007	International Journal of Six Sigma and Competitive Advantage	UK	TPM, Six Sigma	5	Define, measure, analyze, improve, control	All SMEs

(Continued)

Table 6.1 (Continued) Comparison of frameworks

S. No.	Name of author	Year	Journal name	Country	Maintenance framework	No. of pillars	Name of element	Company
14	Ahuja, I. P. S. and Khamba, J. S.	2008	International Journal of Quality & Reliability Management	India	Total Productive Maintenance (TPM)	11	Autonomous, focused, planned, quality, safety, training, TPM, orientation management, loss, zero defects, effectiveness	Manufacturing industry
15	Mishra, R. P., et al.	2008	International Journal of Management Practice	India	TPM	8	Focused improvement, TPM, autonomous maintenance, training, safety, equipment management, quality, planned maintenance	All organizations
16	Anand, G., et al.	2009	International Journal Services and Operations Management	India	Lean manufacturing	9	Elimination of waste, partnership, production, analysis, tools and techniques, management, optimization, 5S, autonomous maintenance	Gear manufacturing
17	Márquez, A. C., et al.	2009	Safety, Reliability and Risk Analysis: Theory Method & Application	Spain	Maintenance management	8	Define, critical analysis, failure cause, TPM, RCM, cost, reliability, risk	Support structure for management
18	Panayiotou, N. A., et al.	2009	International Journal Information Technology & Management	Greece	Business information	4	Situational analysis, strategic analysis, processes, performance measurement	For all organizations
19	Baglee, D. and Knowles, M.	2010	International Journal Control and Cybernetics	UK	TPM	7	Current maintenance, identification, cost, tools, design, strategy change, evaluate	Printing company

(Continued)

Table 6.1 (Continued) Comparison of frameworks

S. No.	Name of author	Year	Journal name	Country	Maintenance framework	No. of pillars	Name of element	Company
20	Jones, E. C., et al.	2010	Total Quality Management and Business Excellence	USA	Six Sigma	9	Commitment, finance, DMAIC (design,measure, analyze, improve and control), execution (4), performance	All companies
21	Muchiri, P.	2010	International Journal of Production Economics	Belgium	Performance indicators	5	Strategy formulation, performance indicators, OEE, benchmarking, performance analysis	Any SME
22	Rose, A. M. N., et al.	2010	Asia Pacific Industrial Engineering and Management Systems Conference	Malaysia	Lean implementation	6	Quality, 5S, preventive maintenance, continuous improvement, KANBAN, customer supplier	Automotive company
23	De Souza, R., et al.	2011	Benchmarking: An International Journal	Singapore	Interview basis	6	Strategic dimensions/goals, social dimension/relation, management dimensions / governance, employee knowledge, technology	Service parts logistics
24	Casals, F. E.	2011	International Conference on E-business, Management and Economics (IPEDR)	UK	Strategic, management and social	5	Service objective and goals, network, technology processes, training, incentive	Any SME
25	Singh, R. K.	2011	Asian Journal on Quality	India	TQM	11	Management commitment, training, participation, coordination, supplier development, feedback, reporting, process, product design, product quality, customer satisfaction	All SMEs

(Continued)

Table 6.1 (Continued) Comparison of frameworks

S. No.	Name of author	Year	Journal name	Country	Maintenance framework	No. of pillars	Name of element	Company
26	Jamian, R., et al.	2012	Asia Pacific Journal of Operations Management	Malaysia	5S	5	Management commitment / participation / training, 5S, Manage service providers, SPs, performance measure, operational performance	All SMEs
27	Nowduri, S.,	2012	World Journal Management	Bloomsburg, USA	Environmental, Quality	7	Environment, social, profit, people, environment, natural economical, sustainability	SMEs
28	Okhovat, M. A., et al.	2012	Scientific Research and Essays	Malaysia	Six Sigma, Lean, TPM	9	DMAIC, WCM, Lean, Six Sigma, leadership / training	Any SME
29	Steyn, A. A., et al.	2012	South African Journal of Entrepreneurship of Small Business Management	Pretoria	Information Communication Technology	6	Data collection, data analysis, interaction, affiliation, environment, identities	Any SME
30	Sharma, R. K., et al.	2013	Quality and Reliability Engineering International	India	Six Sigma TPM integration	10	Strategy, data collection, control planning, improvement, customer satisfaction, define, measure, analyze, improve	Cell of paper manufacturing
31	Jain, A., et al.	2014	International Journal of Lean Six-Sigma	India	TPM	8	Autonomous maintenance, focused maintenance, TPM, quality, training, planned maintenance, safety management	SMEs and large Industries
32	Jie, J. C. R., et al.	2014	International Conference on Industrial Engineering and Operation Management	Bali	Lean Six Sigma	6	Define, measure, analyze, improve, control, Facility condition accessment	Printing company

(Continued)

Table 6.1 (Continued) Comparison of frameworks

S. No.	Name of author	Year	Journal name	Country	Maintenance framework	No. of pillars	Name of element	Company
33	Löfving, M., et al.	2014	Journal of Manufacturing Management	Sweden	Management strategy, SMEs	4	Identification of strategy, tools and techniques, involvement with communication, process	All SMEs
34	Macchi, M., et al.	2014	International Federation for Information Processing	Milano	TPM / Maintenance management	10	Plan, machines, maintenance execution, failure reporting analysis, performance, continuous improvement, HRM, information, opportunity	Manufacturing SMEs
35	Mane, A. M. and Jayadeva, C. T.	2015	International Journal of Process Management & Benchmarking	India	5S	7	5S, leadership, training, teamwork, quality, quality improvement, performances	Indian SMEs
36	Gupta, G., et al.	2015	12th Global Conference on Sustainable Manufacturing	India	WCM / Interpretive Structural Modeling	4	Planning, control (quality, morale, productivity, cost), quality (organize), preparation	All organizations
37	Parikh, Y., et al.	2015	International Journal of Innovative Research in Advanced Engineering	India	Total Productive Maintenance (TPM)/ OEE	11	Autonomous maintenance, focused, planned, quality maintenance, training, safety, TPM, management, zero defects, effectiveness, equipment	Quality of any manufacturing firm
38	Lampadarios, E., et al.	2015	Global Management Perspectives Conference	Italy	Empirical framework of SMEs	3	Entrepreneur factor (age / personality / experience), enterprise factor (capital / product development, strategy, marketing), business environment (economical, cultural)	All SMEs

(Continued)

Table 6.1 (Continued) Comparison of frameworks

S. No.	Name of author	Year	Journal name	Country	Maintenance framework	No. of pillars	Name of element	Company
39	AlManei, M., et al.	2017	ELSEVIER	UK	Lean management	9	Management, training, policy, development, employee, working, communication, finance, customer relations	Cable products
40	Nesamoorthy, R. and Singh, M., et al.	2017	International Journal of Engineering Research and Application	Rajasthan, India	TPM, Six Sigma, Lean, WCM	8	Plan/implementation, measure/identify, lean, continuous improvement, benchmarking, Kaizen, TPM, CAD/CAM / ERP	Indian industries
41	Joshi, K. M., et al.	2018	International Journal of Advanced Research in Engineering & Technology	India	Lean Manufacturing Competitive Scheme	14	Top management, TPM, maintenance definition, rework, information, employee, continuous assessment, defend, unplanned maintenance, major-minor accidents, satisfaction, safety, quality	Textile machineries
42	Sahoo, S. and Yadav, S.	2018	Global Conference on Sustainable Manufacturing	India	TQM	4	Management, strategy planning, involvement, training	All SMEs
43	Yadav, V., et al.	2019	Production, Planning & Control	India	Lean manufacturing	3	5S, KANBAN, processes, SPC, KAIZEN, elimination of waste, production, operation / environmental, financial / social impacts, CSF	All SMEs
44	Mittal, S., et al.	2020	International Journal of Production Research	USA	Smart manufacturing	4	Identification, data assessment, developments, tools identification	SMEs

6.2 SWOT ANALYSIS OF MAINTENANCE FRAMEWORKS

Stanford Research Institute developed a methodology in the beginning that appears to be from the management of enterprises. The background to SWOT originated because of the necessity to discover why enterprises plan unsuccessful. Initially, SWOT was used for enterprises' strengths, weaknesses, opportunities and threats. For evaluating the internal strengths and weaknesses of any enterprise and its outward opportunities and threats internally, a SWOT analysis will be used. It is an important tool for different combinations of planning strategies. To know the enterprise location in the marketplace, it concentrates on the essential components and acts as a precious instrument for SMEs. It is mentioned as the 'school of art and design' (Machmud and Sidharta, 2014), which tries to resolve the problem of various strategies preparation from two expected components – one is from an outside evaluation (in the form of opportunities and threats) and another is from an internal evaluation (in the form of strengths and weaknesses of an enterprise). Both the components are different from each other in terms of the level of confidence achieved. SWOT analysis is primarily used to quantify an enterprise's current strategy and, more particularly, its strengths and weaknesses. Various researchers have utilized SWOT analysis in their framework literature (e.g., Machmud and Sidharta, 2014; Hai, 2008; Kalpande and Gupta, 2015). Another goal relates to the extent of a company's ongoing plan's implementation, as well as how well it deals with changes in the business environment and how well its strengths and weaknesses are tied to those changes. As explained by supporters of SWOT, strengths relate to essential capabilities to participate and become strong. Weaknesses are the essential shortcomings that weaken progress and existence. Both strengths and weaknesses are generally inner side. Opportunities are the better chances and possible vacancies promoting the growth of SMEs. Threats are outward difficulties, which could repress essential capabilities, accelerate weakness and suppress possibilities from existence explodes (Table 6.2).

Depending upon previous concepts, the SWOT analysis defined the Strengths, Weaknesses, Opportunities and Threats of various maintenance frameworks. Because some maintenance frameworks have common structures in which similar strengths, weaknesses, opportunities and threats will be present, it is a logical approach to apply SWOT to the already referred groups. Many frameworks are specific in which the proposals of SWOT analysis are individually given and they are categorized into groups like A, B and C. Unique frameworks are separately collected and analyzed in a

Table 6.2 SWOT definition of different groups made for SWOT analysis

Strength: While comparing maintenance frameworks, if any one of them has a special aspect/characteristic, it is considered as the framework strength.	**Weakness:** When comparing multiple maintenance frameworks, the absence of common parts is seen as a framework weakness.
Opportunity: In maintenance frameworks, if any component that is not required for maintenance execution or is related with maintenance might provide significant comparative benefits to the company, it is considered to provide chances for other frameworks.	**Threat:** Any component of the framework, which is not an essential one for maintenance, could destroy the whole operation; so, it is taken into consideration as a threat. A threat may refer to anything that could cause damage to your organization, venture, or product.

Table 6.3 Groups for the SWOT analysis

Groups	Authors/Consultants	No. of pillars	Remarks
Group A	Labib (1999), Venkatesh (2007), Thomas and Lewis (2007), Ahuja Khamba (2008), Mishra et al. (2008), Baglee and Knowles (2010), Okhovat et al. (2012), Jain et al. (2014), Macchi et al. (2014), Nesamoorthy and Singh (2017) and Jamian et al. (2012), Mane and Jayadeva (2015), etc.	3–10	Quantitative approach
Group B	Kumar et al. (2006), Rose et al. (2010), Okhovat et al. (2012), Almanei et al. (2017), Nesamoorthy and Singh (2017), Joshi (2018), Yadav et al. (2019) and Márquez et al. (2009), Macchi et al. (2014) Löfving et al. (2014), etc.	3–11	Theoretical approach
Group C	Labib (1999), Yusof and Aspinwall (2000), Chang (2005), Singh (2011), Sahoo and Yadav (2018), Kumar et al. (2006), Thomas and Lewis (2007), Jones et al. (2010), Okhovat et al. (2012), Nesamoorthy and Singh (2017), etc.	3–14	Basic definition of TQM
Group D	Mishra et al. (2006), Gupta and Mishra (2016), Nesamoorthy and Singh (2017), Waeyenbergh and Pintelon (2002), Kutucuoglu et al. (2001), Panayiotou et al. (2009), Muchiri et al. (2011), Casals (2010), Nowduri (2012), etc.	3–9	Different approaches from the rest

group named D. Author collected similar kinds of frameworks in the same category of SWOT, grouped and analyzed them on similarities like development year, framework component, nature and type (Table 6.3).

6.2.1   Group A frameworks

This chapter considered 14 frameworks that have been taken in Group A. In this group category, the combination of TPM and 5S has been taken. In Table 6.4, researchers Labib (1999), Venkatesh (2007), Thomas and Lewis (2007), Ahuja and Khamba (2008), Mishra et al. (2008), Baglee and Knowles (2010), Okhovat et al. (2012), Jain et al. (2014), Macchi et al. (2014), Nesamoorthy and Singh (2017) and Jamian et al. (2012), Mane and Jayadeva (2015), etc., are considered.

6.2.2   Group B frameworks

Table 6.5 considered 13 frameworks. Kumar et al. (2006), Rose et al. (2010), Okhovat et al. (2012), Almanei et al. (2017), Nesamoorthy and Singh (2017), Joshi (2018), Yadav et al. (2019) and Márquez et al. (2009), Macchi et al. (2014) Löfving et al. (2014), etc., have been taken in Group B. In this group category, the combination of lean manufacturing, maintenance and personal management has been taken.

Table 6.4 Group A

Strengths	Weaknesses
• Framework accepted widely • Better planning and control of expenses • Continuous improvement • Improve efficiency economically • Facilitate zero defects, failures or accidents • Reduction of waste and increased production • Utilization of resources for quality control • Reduce operation and maintenance cost • Improvement in the team spirit environment • Better cleaning in the working environment • Promote proper documentation • Improve flexibility in working • Reduce maintenance cost of enterprises • Improve existing maintenance practices • Design deficiencies improved in machines	• Higher implementation cost • Not enough appraisal of insecurity and risk • Practical implementations difficult to compare to theoretical charts • Inadequate knowledge of maintenance basics • Insufficient facilitation for employee training • Initiatives in management paradox. • Variance in management initiatives • Defiance in data analysis • Lacking in daily discipline • Inconsistent and unclear expectations • Lack of dedicated representative • Insufficient human resources in terms of expertise/skills • Lack of team spirit between production and non-production like administration, finance
Opportunities	**Threats**
• Better inventory control • Improve job satisfaction for workers • Utilization of floor space for new product lines • Improve safety measures for minimum accidents • Improvisation in product launch time • Development of innovative designs for maintenance prevention • Reduce maintenance cost of machines • Rapid response to market changes • Assist to obtain zero breakdowns	• Increase health issues of workers due to stress • Inconsistent and unclear expectations • Lead to a lack of concentration among employees • For implementation, high skilled workers needed • Resistance from employees • Workers' autonomy decreases

6.2.3   Group C frameworks

This chapter's paper examined 13 frameworks from Group C that are shown in Table 6.6. In this group category, the combination of total quality maintenance (TQM), Benchmarking and Six Sigma has been taken. Labib (1999), Yusof and Aspinwall (2000), Chang (2005), Singh (2011), Sahoo and Yadav (2018), Kumar et al. (2006), Thomas and Lewis (2007), Jones et al. (2010), Okhovat et al. (2012), Nesamoorthy and Singh (2017), etc., are considered.

6.2.4   Group D frameworks

This chapter considered 12 frameworks that have been taken in Group D shown in Table 6.7. In this group category, the combination of WCM, RCM and others

Table 6.5 Group B

Strengths	Weaknesses
• Better utilization of floor space • Qualitative improvement • Improve customer interactions • Efficient reduction in production cost • Waste control during mass production • Maintaining cleanliness in the workplace • Promote inventory control during production • Better consistency of workers • Minimum maintenance cost of the organization • Centralized point-of-contact for suppliers	• Higher installation and maintenance cost • Lack of flexibility for errors • Not enough appraisal of insecurity and risk • Practical implementations difficult to compare to theoretical charts • Inadequate knowledge of maintenance basics • Insufficient facilitation for employee training • Initiatives in management paradox. • Defiance in data analysis • Lack of dedicated representative

Opportunities	Threats
• Time-saving in operations • Increase morale of employees • Transparency of material flows in the company • Improve safety measures, thus minimum accidents • Reduction in time for new product • Achieving zero failures • Better monitoring system • Improve productivity	• Strategic errors made by the Lean movement • Miscalculations that have helped shape what went wrong • Lead to a lack of concentration among employees • Resistance from employees • Less freedom for workers • High cost of initial investment

Table 6.6 Group C

Strengths	Weaknesses
• Improve the quality of product consistently • Better condition for payment deliveries • Increase team spirit among employees • Maintain competitive prices for products • Improve performance with competitors • Reduction in cost of labor expenses • Staff behavior during service to customers • Increase flexibility in working • Convert business strategy into tactics tasks • Improve customer satisfaction • Integrate business-level performance • Proven successful results • Improve employees' efforts toward product quality • Lower warranty and liability costs • Fewer defects and rework rates, less waste	• Insufficient number of trained persons • Underutilization of capacity • Strong competition in the market • Increase in employees absenteeism • Insufficient inventory storage • Management contradiction • Multiple taxes problem in organizations • Conflict in the analysis of the data • Lack of quality consciousness • Lack of proper work culture • Lack of financial strength • Lack of long-term strategic focus • Inadequate attention to research and development

(Continued)

Table 6.6 (Continued) Group C

Opportunities	Threats
• Excellent customer service • Increasing the products portfolio • Possibility of collaborating with new partners • Could move into expert market • Partnership with other organizations regarding the import of some products • Reservation of product item by government • Rapid response to market changes • Achieving zero failures • Obtaining excise relief • Improve flexibility • Increase competitiveness	• Competition from large multinationals • Employee health problem • Increase in the price of inputs • Fluctuating and doubtful prospects • Financial strictness • Technological obsolesces • Negligence toward industrial training • Resistance from employees • Lack of freedom for workers • Lack of political peace and stability

Table 6.7 Group D

Strengths	Weaknesses
• Appropriate data analysis and pickups data • Better coordinate with various duties • Improve inventory control management • Quality processes for systematic work order • Optimize plan and schedule processes • Computerized maintenance management system (CMMS) helps improve maintaining cleanliness • Working with practices and system • Multi-skilled and high trained operator • Improve team spirit among employees • Increase system reliability by failure analysis • Reduce maintenance costs effectively	• Higher investment cost due to computers used • High initial cost leading to revenue loss • Expertise required for implementation • Practical implementations are difficult to compare to theoretical charts. • Lack of quantitative reliability analysis • Lack of in-house training facilities • Initiatives in management paradox. • Defiance in data analysis • Lacking in daily discipline • Confined appraisal of threats and failures • Long time required for implementation
Opportunities	**Threats**
• Profit maximization in various companies • Continuous improvement in the framework • Higher production rate • Involve in continuous improvement • Provide intra-organizational facility • Integration enterprises planning • Increase safety and minimize accidents • Reduce item/equipment replacement	• Data security problem • Problem of secret particulars leaked by the advisor • Non-fulfillment of standards like ISO 9000 • Implementation consultant required • High cost of initial investment • Reconstruction and redesign required • Resistance from employees

(i.e., performance measures, business information, performance indicators, supply chain, information technology, smart manufacturing, etc.) has been taken. Mishra et al. (2006), Gupta and Mishra (2016), Nesamoorthy and Singh (2017), Waeyenbergh and Pintelon (2002), Kutucuoglu et al. (2001), Panayiotou et al. (2009), Muchiri et al. (2011), Casals (2010), Nowduri (2012), etc., are considered.

6.3 CONCLUSION

This chapter presented a brief history of 'framework' to understand how various maintenance practices/systems have developed with respect to time. Currently, maintenance has been extensively used and accepted by various SMEs for different causes. Consequently, the review of the literature on maintenance disclosed that many frameworks are required for the implementation of maintenance in SMEs. These frameworks are different for different enterprises which facilitates us to the conclusion that every enterprise has its own specific framework, although their aims could be nearly the same. The authors considered 44 maintenance frameworks which were initially designed and developed by academicians, consultants and research scholars of various countries and accepted by almost all types of SMEs.

During implementation, it is a tough task for managers or scholars to choose specifically suitable maintenance frameworks for SMEs. By comparing all the frameworks, it was observed that very few are remarkable, although generally in frameworks number of elements/ pillars and their names are slightly different from each other. So, implementation and selection of maintenance framework is a significant decision with zero possibility of mistakes. Hence, we are using a decision-making tool named SWOT analysis for analyzing the various aspects of maintenance framework known as strengths, weaknesses, opportunities and threats. SWOT analysis also recommended that maintenance implementation has not been easy work because it is massively weighted with weaknesses, threats, etc. If organizations worked on the implementation phase, it also generates considerable strengths and opportunities to fulfill the benefit of competition.

A comparative study of different maintenance frameworks disclosed that the strategies followed by various frameworks can be utilized to categorize them into four separate groups – A, B, C and D.

Frameworks of Group A engaged quantitative approaches, Group B engaged conceptual approaches, Group C engaged basic TQM approaches and Group D engaged various other approaches of maintenance used in small and medium scale industries.

The conclusions of each framework group are mentioned below:

- **Group A:** Frameworks could be useful for providing the relation between maintenance efforts and reliability of the system, although these are very tedious, have lots of requirements of data input and are very much complex in nature.
- **Group B:** Frameworks are important for establishing the conceptual classification of maintenance ideas linked to planning, preventive maintenance, and continuous improvement. Frameworks in this category provide an adequate means to pick the suitable maintenance strategy while reducing framework maintenance costs.

- **Group C:** Frameworks are useful in the maintenance practices of various industries which are based on quality-based analysis. Qualitative analysis is based on computer-aided quality control or total quality maintenance frameworks. This group also has some lacking reliable quantity analysis similar to Group B.
- **Group D:** These frameworks each have their own philosophy and methodology.. These are not extensively used approaches as adopted by previous categorized groups but could be adopted by various SMEs according to the requirement of maintenance frameworks. These are less reliable but could be useful because of their uniqueness.

The awareness of the strengths and weaknesses of various strategies represented in the research paper can help business experts take a decision among various approaches depending on the necessities of the SMEs and available resources.

REFERENCES

Ahuja, I. P. S. and Khamba, J. S. (2008). Total productive maintenance: Literature review and directions, *International Journal of Quality and Reliability Management*, Vol. 25, No. 7, pp. 709–756.

Almanei, M., Salonitis, K. and Xu, Y. (2017). Lean implementation frameworks: The challenges for SMEs, *Procedia CIRP*. The Author(s), Vol. 25, pp. 750–755.

Authors, F. (2016). Journal of quality in maintenance engineering article information: Human factors in maintenance: A review list of abbreviations, *Journal of Quality in Maintenance Engineering*, Vol. 22, No. 3, pp. 14–25.

Baglee, D. and Knowles, M. (2010). Maintenance strategy development within SMEs, The development of an integrated approach, *Control and Cybernetics*, Vol. 39, No. 1, pp. 275–303.

Casals, F. E. (2010). The SME co-operation framework: A multi-method secondary research approach to SME collaboration, USA, *2010 International Conference on E-business, Management and Economics (IPEDR)*, Vol. 3, pp. 118–124.

Chang, H. H. (2005). The influence of continuous improvement and performance factors in total quality organization, *Total Quality Management & Business Excellence*, Vol. 16, No. 3, pp. 413–437.

Gupta, G. and Mishra, R. P. (2016). A SWOT analysis of reliability centered maintenance framework, *Journal of Quality in Maintenance Engineering*, Vol. 22, No. 2, pp. 130–145.

Hai, H. L. (2008). Assessing the SME's competitive strategies on the impact of environmental factors: A quantitative SWOT analysis application, *WSEAS Transactions on Information Science and Applications*, Vol. 5, No. 12, pp. 1701–1710.

Islam, M. A., Tedford, J. D. and Haemmerle, E. (2006). Strategic risk management approach for small and medium-sized manufacturing enterprises SMEs - A theoretical framework, *ICMIT 2006 Proceedings -2006 IEEE International Conference on Management of Innovation and Technology*, Singapore, Vol. 2, pp. 694–698.

Jain, A., Bhatti, R. and Singh, H. (2014). Total productive maintenance (TPM) implementation practice: A literature review and directions, *International Journal of Lean Six Sigma*, Vol. 5, No. 3, pp. 293–324.

Jamian, R., Ab Rahman, M. N., Deros, B. M. and Ismail, N. Z. N. (2012). A conceptual model towards sustainable management system based upon 5S practice for manufacturing SMEs, *Asia Pacific Journal of Operations Management*, Vol. 1, No. 1, pp. 19–31.

Jones, E. C., Parast, M. M. and Adams, S. G. (2010). A framework for effective Six Sigma implementation, *Total Quality Management and Business Excellence*, Vol. 21, No. 4, pp. 415–424.

Joshi, K. M. (2018). A modified TPM framework for Indian SMEs, *International Journal of Advanced Research in Engineering and Technology*, Vol. 9, No. 6, pp. 1–14.

Kalpande, S. D. and Gupta, R. C. (2015). Study of SMEs for TQM implementation- SWOT analysis, *International Journal of Engineering and Industrial Management*, Vol. 2, pp. 167–177.

Kumar Sharma, R. and Gopal Sharma, R. (2014). Integrating Six Sigma culture and TPM framework to improve manufacturing performance in SMEs, *Quality and Reliability Engineering International*, Vol. 30, No. 5, pp. 745–765.

Kumar, M. et al. (2006). Implementing the lean sigma framework in an Indian SME, A case study, *Production Planning and Control*, Vol. 17, No. 4, pp. 407–423.

Kutucuoglu, K. Y. et al. (2001). A framework for managing maintenance using performance measurement system, *International Journal of Operations and Production Management*, Vol. 21, No. 1, pp. 173–194.

Labib, A. W. (1999). A framework for benchmarking appropriate productive maintenance, *Management Decision*, Volume 37, pp. 792–799.

Löfving, M., Säfsten, K. and Winroth, M. (2014). Manufacturing strategy frameworks suitable for SMEs, *Journal of Manufacturing Technology Management*, Vol. 25, No. 1, pp. 7–26.

Macchi, M., Pozzetti, A. and Fumagalli, L. (2014). Exploring the integration of maintenance with production management in SMEs, *IFIP Advances in Information and Communication Technology*, Volume 438, pp. 507–514.

Machmud, S. and Sidharta, I. (2014). Business models for SMEs in Bandung: SWOT analysis, *Journal Ekonomi, Bisnis & Entrepreneurship*, Vol. 8, No. 1, pp. 51–61.

Mane, A. M. and Jayadeva, C. T. (2015). 5S implementation in Indian SME: A case study, *International Journal of Process Management and Benchmarking*, Vol. 5, No. 4, pp. 483–498.

Márquez, A. C. et al. (2009). The maintenance management framework: A practical view to maintenance management, *Safety, Reliability and Risk Analysis: Theory, Methods and Applications- Proceedings of the Joint ESREL and SRA-Europe Conference*, Vol. 1, pp. 69–674.

Mishra, R. P., Anand, G. and Kodali, R. (2006). Development of a framework for World-class maintenance, *Journal of Advanced Manufacturing Systems*, Vol. 5, No. 2, pp. 141–165.

Mishra, R. P., Anand, G. and Kodali, R. (2008). A SWOT analysis of total productive maintenance frameworks, *International Journal of Management Practice*, Vol. 3, No. 1, pp. 51–81.

Mittal, S. et al. (2020). A smart manufacturing adoption framework for SMEs, *International Journal of Production Research*, Vol. 58, No. 5, pp. 1555–1573.

Muchiri, P. et al. (2011). Development of maintenance function performance measurement framework and indicators, *International Journal of Production Economics*, Vol. 131, No. 1, pp. 295–302.

Nesamoorthy, R. and Singh, M. (2017). Best manufacturing practice adoptions by Indian industries, *International Journal of Engineering Research and Applications*, pp. 1–10.

Nowduri, S. (2012). Framework for sustainability entrepreneurship for small and medium enterprises (SMEs) in an Emerging Economy, *World Journal of Management*, Vol. 4, No. 1, pp. 51–66.

Okhovat, M. A. et al. (2012). Development of world class manufacturing framework by using six-sigma, total productive maintenance and lean, *Scientific Research and Essays*, Vol. 7, No. 50, pp. 4230–4241.

Panayiotou, N. A., Ponis, S. T. and Gayialis, S. P. (2009). Designing an industrial maintenance system: A proposed methodological framework, *International Journal of Information Technology and Management*, Vol. 8, No. 4, pp. 361–381.

Rose, A. M. N., Deros, B. M. and Rahman, M. A. (2010). Development of framework for lean manufacturing implementation in SMEs, *The 11th Asia Pacific Industrial Engineering and Management Systems Conference the 14th Asia Pacific Regional Meeting of International Foundation for Production Research*, Melaka, (December), pp. 1–5.

Sahoo, S. and Yadav, S. (2018). Total quality management in Indian manufacturing SMEs, *Procedia Manufacturing*, Vol. 21, pp. 541–548.

Singh, R. K. (2011). Analyzing the interaction of factors for success of total quality management in SMEs, *Asian Journal on Quality*, Vol. 12, No. 1, pp. 6–19.

Thomas, A. and Lewis, G. (2007). Developing an SME-based integrated TPM – Six Sigma strategy, *International Journal of Six Sigma and Competitive Advantage*, Vol. 3, No. 3, pp. 228–247.

Venkatesh, J. (2007). An introduction to total productive maintenance (TPM), *The Plant Maintenance Resource Center*, pp. 3–20.

Waeyenbergh, G. and Pintelon, L. (2002). A framework for maintenance concept development, *International Journal of Production Economics*, Vol. 77, No. 3, pp. 299–313.

Yadav, V. et al. (2019). The propagation of lean thinking in SMEs, *Production Planning and Control*, Vol. 30, No. 10–12, pp. 854–865.

Yusof, S. M. and Aspinwall, E. (2000), Conceptual framework for TQM implementation for SMEs, *TQM Magazine*, Vol. 12, No. 1, pp. 31–36.

Chapter 7

Development of maintenance framework for SMEs by an ISM approach

Gouraw Beohar, Rajesh P. Mishra, and Anjaney Pandey

CONTENTS

7.1 INTRODUCTION

The smooth running of an SME and optimum utilization of resources at their best capacity is the biggest art in the present highly stressed and volatile environment for
business. Maintenance plays a very important role in improving cost-effectiveness,
quality products, and customer satisfaction during its operation. Quality, availability, and productivity have become a crucial part of manufacturing industries now.
Due to increase in demands varies, complexity and capital investment are major parameters for machines. Framework development and its implementation is a difficult process
because it suffers from the absence of a systematical and uniform methodology. Main reason related to handling the maintenance implementation issues is that all SMEs have their
own different ideas, approaches, and methods. Academicians and researchers of different

DOI: 10.1201/9781003293576-7

countries proposed various maintenance framework models based on several best maintenance practices for entrepreneurs and claimed that such models are suitable for total
maintenance improvement in SMEs. This helps to give motivation to the SMEs established
worldwide for accepted impressive and beneficial maintenance policies such as TQM (Total
Quality Maintenance), TPM (Total Productive Maintenance, LM (Lean Manufacturing),
CBM (Condition-based Maintenance), RCM (Reliability-centered Maintenance), etc.,
over the conventional fire-fighting strategies of maintenance like Breakdown Maintenance
by Sharma (2008). The maintenance framework purposes, implementing parameters, time
of implementation, benefits inside SMEs, and other aspects are explained by Medori and
Steeple (2000). The framework development initiates the identification of the valid existent
system of measurement and it is a proven method to improve measurement systems and
increases the overall performance of the system of SMEs. Yusof and Aspinwall (2000)
utilized the approach of TQM for defining the proposed framework for SMEs. Author
explained that modification, adaptation, and rearrangements in SMEs could be due to
their smaller size, lacking expertise, pressure of target completion, financial problems,
and human resources issues. Kutucuoglu et al. (2001) defined the requirement of a perfect
PMS (Performance Measurement System) in small enterprises. He clarified the typical
approach of performance measurement, which heavily relies on financial and accounting
data and may obstruct or impede the adoption of a PMS.

7.2   FRAMEWORK DESIGN REQUIREMENT

The definition, objectives, needs, purposes, and importance of the framework are already discussed in the previous research section by the author. This paper makes an
effort to concentrate on the framework development and the procedures involved.
A question initially arises as to what are the development procedures for maintenance
framework and how we can describe a good maintenance framework that is suitable for
various types of SMEs. The following are the features of a good maintenance framework:

• Easily understandable and systematic
• Simple structure
• Provides clarity between different steps used and defined elements
• Broad enough to match various circumstances
• Presents a sufficient guideline for plan preparation and its implementation
• Easy implementation procedure, time-saving and cost-effective

Generally, academicians consider some standard criteria which act as a guide in better
framework development and suit SMEs' characteristics (Yusof and Aspinwall, 2000).
A question initially arises about how we can describe a good maintenance framework
that is suitable for various types of SMEs and what its development procedures are.
A conceptual framework was proposed by Deros et al. (2006) for implementing the
benchmarking approach in SMEs in consideration of their qualities. This benchmarking approach helps the SMEs to achieve a high rate of customer satisfaction, effective
business processes, better financial results, competition, and advanced and dedicated
human resources. Two broadly recognized strategies of business process improvements,

i.e., Lean Manufacturing and Six Sigma are presented by Kumar et al. (2006). These are two different approaches presented by author to apply their combined effects on SMEs known as the Lean Sigma framework. It fulfills the overall goal of the enterprises both ways. On one side, product defects, cost of inventory, scraps, rejections, and rework costs are reduced by Lean; on the other side, the use of statistical techniques and tools of Six Sigma result in improved customer satisfaction and performance of SMEs.

In another practice of maintenance, Mishra et al. (2006) proposed a framework that is also called world-class maintenance. This model helped to provide the optimum solution for TPM's shortcomings. Joshi and Bhatt (2018) presented a framework for establishing Lean manufacturing in SMEs and improving productivity, quality, and Overall Equipment Effectiveness (OEE) based on TPM. Smart manufacturing is an innovative approach to increasing output rate that improves SME performance in terms of time, cost, and product quality.Mittal et al. (2019) presented a framework for the integration of technology data with various manufacturing approaches, processes, and technologies.

The latest approach for SMEs is related to a new framework called SCOPE. This is basically to present the model development for framework strategies, challenges, opportunities, and problems. Some valid reasons for maintenance framework development are as follows:

- To demonstrate an outline to discover a new concept for enterprises
- To drive the management's attention to a significant list of major issues that could not be discussed
- To provide information on the enterprise's strengths and weaknesses
- To provide a clear vision for the strengths and weaknesses of SMEs
- To successfully implement transformation processes at the time of process modification
- To help clear a pictorial view for management for competing with the open global market

7.3   DEVELOPMENT PHILOSOPHY OF MAINTENANCE FRAMEWORK

The development of new proposed framework is a tedious job. Deep knowledge and understanding of the existing framework are required before starting any work. It is essential to realize which areas are properly discussed and which areas have not been discussed yet. Author initially considered the comparisons among all existing frameworks used in SMEs which helped and guided him to propose a new maintenance framework. The proposed maintenance framework is truly based on literature review, expert consultation, and area of knowledge about all elements of SMEs' maintenance system. The proposed framework not only provides the best suitable solution for professionals allured mostly from the previous experiences of small enterprises but also overcomes the gaps, failures, and problems considered in published literature and case studies. Consideration of best processes for SME maintenance can be referred to as those maintenance practices that increased the efficiency of SMEs for achieving the benchmarks over their competitors extensively. Development of maintenance framework system includes different sub-elements / activities / attributes and practices such as control of inventory, quality, finance, Human Resorce Department (HRD), etc.

7.3.1   Various elements of the proposed maintenance framework for SMEs

A complete list of various elements, their description, activities, and policies played an important role in the proposed maintenance framework implementation. Established elements are based on the existing maintenance framework, opinions of experts, and best maintenance practices implemented by the various enterprises. Such information may assist or encourage enterprises to use these practices to improve maintenance efforts and overall production performance. In his previous published paper, author compared 43 different existing frameworks used in SMEs.

The comparison table is based on the number of elements used for framework construction by various researchers in their proposed methodologies.

The reason for comparative analysis related to framework development includes various required initiatives (i.e., elements / attributes / variables) that could make a pathway for the attainment of goals.

Approximately, 153 unique elements were noted by comparison in Table 7.1 considered. These (elements / attributes / variables) presented the various pillars of frameworks.

The main problem is related to the adaptation of all these 153 elements in a fragmented manner because there are few relations that exist between all these elements. Identification of elements grouped them into a key element and defined the relation between them helping in the development of a new framework which may be provided ways to achieve total maintenance in an SME's system. After discussion with various experts, practitioners, researchers, and consultants, the output of the comparative analysis was grouped in the best possible way and utilized by author for future proposed framework model. By the end of the discussion, the author has reorganised 153 distinct pieces into 59 unified (key) elements that are comparable, synonymous, and synchronised.

7.3.2   Development of key elements with sub-elements for the proposed framework for SMEs

With the help of a superficial review of literature and by consideration of various elements/sub-elements, a maintenance framework could be proposed. Unique (key) attributes/elements can be identified in separate groups and then classified on the basis of common factors moving among themselves (Table 7.2).

These 59 different sub-elements of various maintenance frameworks will rearrange and combine to generate the key attributes/elements in the group of eight different pillars of the proposed framework, as presented in Table 7.3.

To achieve this, an ISM (Interpretive Structural Modeling) approach is used. It is an advanced methodology of interactional planning which permits a group of persons to work as a team and to develop a framework that enforces order and direction on the intricate relationships among a set of elements.

Table 7.1 Various elements of framework

Leadership and management (1)	Ownership maintenance\ Continuous improvement of process and procedure (2)	Human resource development (3)	Lean tools (4)	Quality management (5)	Safety, health, and environment systems (6)	Planned maintenance (7)	Support systems for maintenance improvement (8)
Top management involvement / Leadership / Decisions	Autonomous maintenance / ownership	Motivation / Moral / Employee satisfaction / Empowerment	Elimination of waste / Lean manufacturing	TQM / Quality initiatives / Quality assurance / Quality acceptance / Quality Function Deployment measurement / Zero defects	Work environment/ Culture organization	Maintenance / Plan / Unplanned maintenance / Preventive maintenance equipment	Data collection / Data classification measurement / Critical data analysis / Economical input data
Management commitment / Management practice / SME commitment	Kobetsu Kaizen / Continuous analysis / Continuous improvement	People	Review / Improve / Trial / Standardize	Reliability centered maintenance / World-class maintenance	Unplanned safety health and environment / Safety	Maintenance decision	Definition of problem / Define / Identification of information / Tool identification
Critical Success Factor (CSF)	TPM	Skill / Trained / Training / Utilization Teamwork Multi-skilled workforce	Lesson learn / Mistake proof exercise	Sustainability / Benchmarking	Tradition	Processes / Activity plan / Optimization of maintenance Strategy analysis / Select policy / Maintenance policy	Control chart / Risk control
Profit / Economic / Finance / Cost / Cost of quality		Productivity / Production / Performance indicators (PI)	5 S / Tools	Rework / Customer satisfaction / Reliability / Reliability analysis			Risk management / Risk analysis / Situational analysis

(Continued)

Table 7.1 (Continued) Various elements of framework

Leadership and management (1)	Ownership maintenance\Continuous improvement of process and procedure (2)	Human resource development (3)	Lean tools (4)	Quality management (5)	Safety, health, and environment systems (6)	Planned maintenance (7)	Support systems for maintenance improvement (8)
Basic / Result	OEE / Effectiveness	Appraisals		Product / Product quality / Six Sigma		Awareness / Planning analysis	Failure analysis / Equipment failure / FMEA / FRCA
Entrepreneurial factors / Enterprise factors		Flexibility / Acceptance		Customer service objective / Supplier-customer relation		Office TPM	Maintenance cost analysis / Performance measurement
Development / Tailored vision		Social input / Coordination at company				Select machine / Apply policy to machine / Schedule	Communication channels / Feedback / Customer feedback
Business environment						Opportunity	Delivery / Just In Time (JIT) / Enterprise Resource Planing
General methodology / Institutional philosophy						Stop / Breakdown / Delays	
Refocusing framework /							KANBAN / Supply chain network
Impact models /							
System analysis framework							Smart manufacturing CAD / CAM / CIM / FMS / AIMMS

Table 7.2 Element combine table

Elements under each implementation stage of the proposed framework

S. no. elements / sub-elements / approaches / practices / activities

1. Leadership and management	**5. Quality management**
Management commitment	Quality assurance
Management practice	Reliability
Critical Success Factor (CSF)	Sustainability
Economic activities	Customer satisfaction
Result orientations	Six Sigma
Enterprise factors	Supplier-customer relation
Tailored vision	Benchmarking
	6. Safety, health, and
Business environment	**environment systems**
	Work environment
Institutional philosophy	Unplanned safety
Impact models	
2. Ownership maintenance / continuous	Risk control
improvement of process	
Autonomous maintenance	**7. Planned maintenance**
Continuous improvement	Preventive maintenance
Productivity	Maintenance decision
OEE/Effectiveness	Optimization of maintenance
3. Human resource development	Maintenance policy
Employee satisfaction	Planning analysis
People	Schedule
Education and training	Opportunity
Multi-skilled workforce	Delays
	8. Support systems for
Appraisals	**maintenance improvement**
	Critical data analysis
Flexibility/Acceptance	Identification of information
Organization culture	Situational analysis
4. Lean tools	Failure analysis
Muda / mura / muri	Performance measurement
Standardization	Supply chain network
5S	ERP/CMMS (computerized
Poka-yoke	maintenance management system)
	Office TPM

Table 7.3 Pillars of the proposed framework

S. no.	Various pillars of framework
1	Leadership and management
2	Ownership maintenance / continuous improvement of process and procedure
3	Human resource development
4	Lean tools
5	Quality management
6	Safety, health, and environment systems
7	Planned maintenance
8	Support systems for maintenance improvement

7.4   ISM

ISM was invented by John N. Warfield in the period 1971–1973 to examine complex problems and organize them in terms of phrases and directional graphs that can be easily understood as a computer-aided methodology. ISM technique, based on a computer-assisted mathematical process, helped to realize the complex issues of structure modeling. The process is related to relational mathematics which transforms undefined, unclear, insignificant, wrongly articulated, mental model into a cleared interrelated well-defined structure subsequently transitivity. This is an important tool for maintenance framework with help of some experts from the service, manufacturing, and education sectors. ISM methodology is to be used to follow some clearly identified steps in a specified order as considered. Every step is related to the previous one and could not be ignored. This paper draws up a list that includes each and every step expressly as exactly as possible to help academicians and researchers to understand ISM methodology properly. There are several steps required for applying ISM methodologies.

7.4.1   Problem identification

It is a complicated issue to verify the ISM use for problem-solving because every problem-related issue has sufficient variables for the human brain to consider its complex nature. After analyzing the problem, a person is overwhelmed by a large number of interrelationships and mutual connections among all the variables, which helps to establish a systematical analysis in a readable and comprehensible format.

If it is so, issues are the perfect candidacies with the ISM approach. Development of a maintenance framework is also a complex work for SMEs; so, the adaptation of ISM methodology will be a better choice.

7.4.2   Elements identifications

As a decision has been taken regarding various issues of framework development, the very first step is to identify all the elements involved in the problem. Accurate identification of all the elements could be done by a group of experts, knowledgeable, and experienced persons having the capability to deal with related issues by finding out solutions with methods, e.g., questionnaires, survey methods, etc.

7.4.3   Structure-based decision

Now as we have the relation between the issues and elements, the next step is to define the contextual relationship among all the elements in the model proposed, i.e., trying to find out the relationship of all the available elements. Warfield classified the chosen framework as the priority structure. In framework, one element helps to achieve the other element, e.g., management decision would help to increase the motivation of employees, priority structure directs elements having higher importance/preference over other elements.

7.4.4   To Identify pair-wise relation

Warfield defined the ISM approach to develop a hierarchical model, based on pair-wise relations between elements. This step is related to defining the contextual relationship that exists between elements, i.e., each element will be cross-checked with all other elements to verify the relationship with particular pair. It should be pointed out that contextual relationship establishes directionally – either one way or both ways.

7.4.5   Development of SSIM (Self-Structural Interpretive Matrix)

This step is related to the identification of elements to build in a matrix form. Matrix involves eight elements which are written in the form of Y-axis and X-axis of the matrix. The matrix defines pair-wise relationship between the element on Y-axis and the element on X-axis (Table 7.4).
    Let us define the relationship in the following way:

V: defined forward relation from i to j helps to influence, i.e., $i = 1, j = 0$.
A: defined reverse relation from j to i helps to influence, i.e., $j = 1, i = 0$.
X: defined directional relation both ways, i and j influence each other, i.e., $i = 1, j = 1$.
O: defined that no relation exists between them, i.e., $i = 0, j = 0$.

7.4.6   Initial reachability matrix development

The name "reachability matrix" indicated how many entries in this matrix are accessible from the remaining elements. To show the pair-wise relation in the binary form of 1 and 0, the SSIM matrix is to be converted into a reachability matrix. In this type of matrix, the relationships are to be converted into binary form by applying the conversion given below in Table 7.5.

Table 7.4 SSIM matrix

S. no.	Key elements	1	2	3	4	5	6	7	8
1	Lean tools	X	A	A	A	X	A	A	O
2	Quality management		X	A	A	O	O	V	X
3	Human resource development			X	A	V	X	V	V
4	Leadership culture and commitment				X	V	V	V	V
5	Safe and healthy environment					X	A	X	A
6	Planned maintenance						X	V	O
7	Support system improvement							X	A
8	Ownership maintenance								X

Table 7.5 Initial reachability matrix

S. no.	Key elements	I	2	3	4	5	6	7	8
I	Lean tools	I	0	0	0	I	0	0	0
2	Quality management	I	I	0	0	0	0	I	I
3	Human resource development	I	I	I	0	I	I	I	I
4	Leadership culture and commitment	I	I	I	I	I	I	I	I
5	Safe and healthy environment	I	0	0	0	I	0	I	0
6	Planned maintenance	I	0	0	0	I	I	I	0
7	Support system improvement	I	0	0	0	I	0	I	0
8	Ownership maintenance	0	I	0	0	I	0	I	I

7.4.7 Incorporating transitivity and developing final reachability matrix

Now in the initial reachability matrix, further transitivity is to be checked if any. Transitivity is a fundamental assumption of the contextual relationship of the ISM approach which states that if the 1'st element is related to 2'nd and 2'nd is related to 3'rd, then 1'st will be necessarily related to the 3'rd element. If any gap is found during SSIM development, it would be filled by 1* entries incorporating transitivity. After completion of the previous steps, final reachability matrix is obtained, presented in Table 7.6. Row-wise counting of variables' frequency indicates their driving power, and column-wise counting of frequency indicates their dependent power.

7.4.8 Partitioning of reachability and antecedent sets

A reachability set is defined for every element as a set including elements that can be reachable from the specific element. The set of each element consists of elements whose cells in the row related to the element are assigned "1" in the reachability matrix and so on for the remaining elements.

An antecedent set is defined for each element as a set including elements which could reach that specific element. In other words, it is a set for every element including elements whereof cells inside the column related to the elements are allotted "1" in reachability matrix. Elements that are the same in reachability form and sets of

Table 7.6 Final reachability matrix

S.no.	Elements	I	2	3	4	5	6	7	8	Driving power
I	Lean tools	I	0	0	0	I	0	I*	0	3
2	Quality management	I	I	0	0	I*	0	I	I	5
3	Human resource development	I	I	I	0	I	I	I	I	7
4	Leadership culture and commitment	I	I	I	I	I	I	I	I	8
5	Safe and healthy environment	I	0	0	0	I	0	I	0	3
6	Planned maintenance	I	I*	I	0	I	I	I	I*	7
7	Support system improvement	I	0	0	0	I	0	I	0	3
8	Ownership maintenance	I*	I	0	0	I	0	I	I	5
	Dependent power	8	5	3	I	8	3	8	5	41/41

Table 7.7 Reachability, antecedent, and intersection for iteration I at level I

S. no.	Elements	Reachability	Antecedent	Intersection set	Level
I	Lean tools	I,5,7	I,2,3,4,5,6,7,8	I,5,7	I
2	Quality management	I,2,5,7,8	2,3,4,6,8	2,8	
3	Human resource development	I,2,3,5,6,7,8	3,4,6	3,6	
4	Leadership culture and commitment	I,2,3,4,5,6,7,8	4	4	
5	Safe and healthy environment	I,5,7	I,2,3,4,5,6,7,8	I,5,7	I
6	Planned maintenance	I,2,3,5,6,7,8	3,4,6	3,6	
7	Support system improvement	I,5,7	I,2,3,4,5,6,7,8	I,5,7	I
8	Ownership maintenance	I,2,5,7,8	2,3,4,6,8	2,8	

intersection occupied the highest level in the ISM hierarchy. The process carried on up to all the levels like level 1, level 2, level 3, etc., of elements established (Table 7.7).

Table 7.8 represents the same procedure among the reachability, antecedent, and intersection sets also called iteration 2 at level 2. Process repeat up to all level iterations will not be completed like level 2, level 3, and so on (Tables 7.9 and 7.10).

Table 7.8 Reachability, antecedent, and intersection for iteration 2 at level 2

S. no.	Elements	Reachability	Antecedent	Intersection set	Level
2	Quality management	2,8	2,3,4,6,8	2,8	2
3	Human resource development	2,3,6,8	3,4,6	3,6	
4	Leadership culture and commitment	2,3,4,6,8	4	4	
6	Planned maintenance	2,3,6,8	3,4,6	3,6	
8	Ownership maintenance	2,8	2,3,4,6,8	2,8	2

Table 7.9 Reachability, antecedent, and intersection for iteration 2 at level 2

S. no.	Elements	Reachability	Antecedent	Intersection set	Level
3	Human resource development	3,6	3,4,6	3,6	3
4	Leadership culture and commitment	3,4,6	4	4	
6	Planned maintenance	3,6	3,4,6	3,6	3

Table 7.10 Reachability, antecedent, and intersection for iteration 2 at level 2

S. no.	Elements	Reachability	Antecedent	Intersection set	Level
4	Leadership culture and commitment	4	4	4	4

Table 7.11 Element-level interactions table from Table 7.4 to Table 7.10

S. no.	Key elements	Reachability	Antecedent	Intersection set	Level
1	Lean tools	1,5,7	1,2,3,4,5,6,7,8	1,5,7	I
5	Safe and healthy environment	1,5,7	1,2,3,4,5,6,7,8	1,5,7	I
7	Support system improvement	1,5,7	1,2,3,4,5,6,7,8	1,5,7	I
2	Quality management	1,2,5,7,8	2,3,4,6,8	2,8	II
8	Ownership maintenance	1,2,5,7,8	2,3,4,6,8	2,8	II
3	Human resource development	3,6	3,4,6	3,6	III
6	Planned maintenance	1,2,3,5,6,7,8	3,4,6	3,6	III
4	Leadership culture and commitment	1,2,3,4,5,6,7,8	4	4	IV

Table 7.12 Conical matrix

S. no.	Key elements	1	5	7	2	8	3	6	4	Driving power
1	Lean tools	1	1	1	0	0	0	0	0	3
5	Safe and healthy environment	1	1	1	0	0	0	0	0	3
7	Support system improvement	1	1	1	0	0	0	0	0	3
2	Quality management	1	1	1	1	1	0	0	0	5
8	Ownership maintenance	1	1	1	1	1	0	0	0	5
3	Human resource development	1	1	1	1	1	1	1	0	7
6	Planned maintenance	1	1	1	1	1	1	1	0	7
4	Leadership culture and commitment	1	1	1	1	1	1	1	1	8
Dependent power		8	8	8	5	5	3	3	1	

7.4.9   Development of conical matrix

A conical matrix is formed by combined together enablers at the same level, across rows and columns of the final reachability matrix. Before presenting conical matrix, paper discusses a consolidated table on element-level interactions from Table 7.4 to Table 7.10 (Table 7.11).

7.4.10   Digraph development

The generation of digraph is mainly based upon conical form of reachability matrix. It includes transitivity links formed by lines of edges and nodes (Figure 7.1).

Assume that there is some relation existing between two elements, which is demonstrated by an arrow from one element to another element. As hierarchy confirmed, the action is indicated by an arrow direction. If element *i* directs element *j*, an arrow will be pointing out from *i* to *j* direction. The process will be carried out over all the elements until their relations are not confirmed. In the same way, after defining the relationship (directly or indirectly) among all the elements, diagram could be achieved. Finally, the digraph will be developed by the elimination of all the indirect links as shown in Figure 7.1. The elements found in the initial iteration level are placed at 1st position in the digraph and the next level will be placed in 2nd position and so on up to the last/final iteration level placed at the last position in the digraph. The next step is related to the convergent of digraph into an ISM model by nodes replacement of elements with statements. Level categorization of various elements is clearly visible in the final digraph (Figure 7.2).

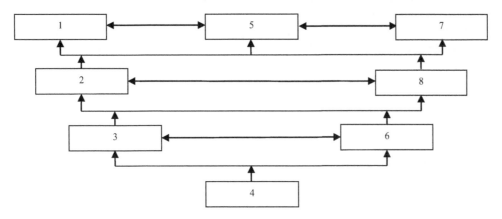

Figure 7.1 Digraph.

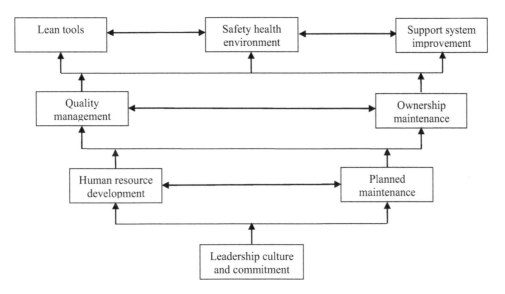

Figure 7.2 Final ISM model.

7.5 CONCLUSION

Finally, the leadership culture and commitment are the most valuable elements of the proposed framework for SMEs. This element was working as a major part of the framework which was essential to fulfill all kinds of maintenance requirements of SMEs. Therefore, the Indian manufacturing sector has to concentrate on these elements to establish a framework effectively and efficiently. Paper especially develops a framework model to research the elements of Indian SMEs which helps to present a common procedure to be applied in different sectors also. Worldwide development leads to a high level of competitiveness among all the SMEs in manufacturing industries for developing higher reliability, availability, and sustainability in their frameworks. Unfortunately, limited knowledge is available concerned with SMEs, especially in developing countries like India. Previously available research literature and studies very limitedly explain the practical implementation problems for maintenance framework development. The originality of the paper arises from the fact that a complete set of elements has been identified and considered to be gathered on the grounds of an in-depth study of literature (Table 7.12).

Development of conceptual framework is restricted to the identification of the elements, expert discussion, relation findings, related output in maintenance, etc. The future scope of the paper could be related to other aspects of SMEs in various sectors also.

REFERENCES

Deros, B., Yusof, S. M., and Salleh, A. M. (2006). A benchmarking implementation framework for automotive manufacturing SMEs. *Benchmarking: An International Journal*, Vol. 13, No. 4, pp. 396–430.

Joshi, K. M., and Bhatt, D. V. (2018). A modified TPM framework for Indian SMEs. *International Journal of Advanced Research in Engineering and Technology*, Vol. 9, No. 6, pp. 1–14.

Kumar, M., Antony, J., Singh, R. K., Tiwari, M. K., and Perry, D. (2006). Implementing the Lean Sigma framework in an Indian SME: A case study. *Production Planning & Control*, Vol. 17, No. 4, pp. 407–423.

Kutucuoglu, K. Y., Hamali, J., Irani, Z., and Sharp, J. M. (2001). A framework for managing maintenance using performance measurement systems. *International Journal of Operations & Production Management*, Vol. 21, No. 1/2, pp. 173–194.

Medori, D., and Steeple, D. (2000). A framework for auditing and enhancing performance measurement systems. *International Journal of Operations & Production Management*, Vol. 20, No. 5, pp. 520–533.

Mishra, R. P., Anand, G., and Kodali, R. (2006). Development of a framework for world-class maintenance systems. *Journal of Advanced Manufacturing Systems*, Vol. 5, No. 2, pp. 141–165.

Mittal, S., Khan, M. A., Purohit, J. K., Menon, K., Romero, D., and Wuest, T. (2019). A smart manufacturing adoption framework for SMEs. *International Journal of Production Research*, Vol. 58, No. 2, pp. 1–19.

Yusof, S. M., and Aspinwall, E. (2000). A conceptual framework for TQM implementation for SMEs. *The TQM Magazine*, Vol. 12, No. 1, pp. 31–36.

Chapter **8**

Multi-objective parametric optimization of wire electric discharge machining for Die Hard Steels using supervised machine learning techniques

Pratyush Bhatt, Pranav Taneja, and Navriti Gupta

CONTENTS

8.1   INTRODUCTION

Hardened steel is produced by applying heat treatment, quenching and subsequent reheating processes on plain carbon steel. Hardened steel possesses resistance to wear, severe abrasion as well as impact and shock forces [1]. Corrosion-resistant coating (generally chromium coating) can be applied to die steels in order to further improve their resistive properties. Components made of hardened steel have a hard exterior casing [2,3]. Some of the components made from hardened steel are arbors, axles, driving pinions and camshafts. Components made from hardened steel find extensive application in fields such as power production, transportation and general manufacturing engineering [4]. These mechanical properties make hardened steel the most widely used material for dies and other cutting tools. But these properties are also

DOI: 10.1201/9781003293576-8

responsible for making hardened steels difficult to machine using conventional machining techniques. The material remains brittle even after tempering and, thus, it sustains damage from sharp impacts. Thus, precision machining of the alloy becomes a great challenge.

WEDM is an unconventional machining process that removes material through repeated electrical discharges of short periods and high current intensity between workpiece (anode) and wire (cathode). Spark discharges are generated between cathode and anode. Deionized water acts as a dielectric medium and erodes workpiece. Numerically Controlled (NC) programmed path enables the production of complex geometrical shapes. The only requirement of WEDM is that tool and workpiece must be made up of electrically conductive material. Mechanical forces are not exerted on the workpiece because it is not in contact with the electrode. Therefore, WEDM can machine tough and brittle materials with precision in a cost-effective manner [4–10]. Hence, WEDM is ideal for machining die hard steels efficiently [6–7]. The various input or response parameters involved in WEDM are explained below:

Pulse ON time (T_{on}): It is the time period during which the electrodes develop potential difference. In this period, sparks are generated by the passage of current between the electrode and workpiece. It is measured in µs [5].

Pulse OFF time (T_{off}): It is the time period between two discharges during which the voltage is absent. It is measured in µs.

Wire feed: It is the speed of wire movement in wire guide path. It is measured in mm/min.

Servo voltage: It is the potential difference between tool wire and workpiece. It is measured in volts. Servo stabilizer is used with EDM (Electrical Discharge Machining) machines to protect it from unbalanced voltage. If there is a voltage fluctuation, these machines may malfunction because the EDM machine comprises several micro parts that act as the controller unit.

The output variables obtained in WEDM are:

Surface Roughness (R_a): It is a measure of irregularities on machined surfaces or the measure of vertical aberrations along a flat surface. It is measured in µm using a surface roughness comparator or profilometer.

Material Removal Rate (MRR): It refers to the material removed per unit time. It is measured in mm^3/min [6].

$$MRR = \text{Kerf width}(k) \times \text{Thickness of workpiece}(t) \times \text{Cutting velocity}(V_c)$$

Experimental determination of the optimum input parameters for desired R_a and MRR in WEDM is an effective and accurate method. But due to the slow rate of MRR, a lot of time is consumed in the process, especially for large-sized workpieces. Also, excessive power is consumed in the WEDM operation. Moreover, large capital is required for the experimentation. Substitution of experimental analysis by supervised machine learning algorithms for the prediction of R_a and MRR resolves these issues. In supervised machine learning, computational or mathematical algorithms are trained using "labeled" training data. Labeled data comprises of a collection of input variables along with their corresponding output variables. This enables the algorithms to predict the output for unseen data having input and output fields similar to the training data. Therefore, a supervised machine learning algorithm aims to generate a mapping function that can accurately match the input variables to the output variables.

In recent years, many kinds of research have been performed to optimize response parameters of the WEDM process for different materials using either the analytical Taguchi method or machine learning techniques. Deshwal et al. [6] studied the impact of multiple response parameters on MRR and R_a using the Taguchi method on H13 die steel. Kumar [7] performed fine-tuning of WEDM response parameters in machining H13 die steel using Grey-Taguchi method. Pragnesh et al. [8] performed single-objective optimization of various input parameters like peak current, T_{on} and T_{off} for desired surface roughness using Response Surface Methodology for WEDM. Ulas et al. [9] predicted R_a of Al7075 alloy machined by WEDM with machine learning techniques like Extreme Learning Machines, Weighted Extreme Learning Machines, Support Vector Regression, Quantum Support Vector Machine. Shukla and Priyadarshini [5] made use of gradient descent machine learning algorithm to optimize surface roughness as well as kerf width of Hastelloy C276 with WEDM. Singh et al. [10] investigated the use of Support Vector Machines (SVM), Gaussian process and Artificial Neural Network (ANN) to evaluate surface roughness using WEDM on Nimonic-90 superalloy. Phate et al. [11] used ANN-assisted Principal Component Analysis (PCA) to optimize multiple response parameters for WEDM on aluminum silicate composite. Most researchers performed single-objective optimization. Gupta et al. [12–14] applied ANN-based predictive models in predicting MRR and surface finish in WEDM and turning process in their research work. Machine learning (ML) techniques have been barely employed to model WEDM for H13 die steels.

The objective of the study is to map the response parameters – servo voltage, T_{on}, T_{off} and wire feed – to surface roughness and MRR using several supervised machine learning models such as LASSO regression, Support Vector Regression, K-nearest neighbors (KNN) regression and Neural Network regression. Thus, optimization of multiple input parameters (Multi-parametric optimization) for desired output parameters – surface roughness and MRR – is performed simultaneously (Multi-objective optimization) for WEDM of H13 die hard steel. The performance of ML models is evaluated using the metric named mean squared error (MSE), and the technique which most accurately models the WEDM process for such a small dataset is identified.

8.2   RESEARCH METHODOLOGY

8.2.1   Dataset description

The dataset was obtained from WEDM experiments on H13 hot die steel from literature [6]. The dataset consists of 16 runs or experiments with 4 levels of 4 input parameters (T_{on}, T_{off}, feed rate and servo voltage) and the corresponding 2 output parameters (R_a and MRR) obtained from the experiments. The response parameters along with their levels are given in Table 8.1.

Table 8.1 Four levels of response parameters

Level	Pulse ON time	Pulse OFF time	Wire feed	Servo voltage
1	105	25	5	35
2	110	35	7	40
3	115	40	9	45
4	120	45	11	50

8.2.2 Supervised machine learning models

8.2.2.1   LASSO regression

Least Absolute Shrinkage and Selection Operator (LASSO) Regression is a form of linear regression that utilizes shrinkage. In shrinkage, the data points are scaled-down toward their mean. The LASSO regularization converts the solution into simple and sparse models by reduction of coefficients to zero and thus reducing the number of input parameters. LASSO regression uses L1 regularization [15] in which the summation of absolute values of coefficients is added as a penalty to the residual sum of squares (loss) after multiplication with the regularization parameter (λ). Large values of λ may completely eliminate most input parameters and therefore need to be set using cross-validation [16]. The objective of the algorithm is to minimize [17]:

$$\sum_{i=1}^{n}\left(y_i - \sum_{j} x_{ij}\beta_j \right)^2 + \lambda\sum_{j=1}^{p} |\beta_j|$$

where λ is a tuning parameter that regulates the effect of L1 penalty [18]. When a few βs are minimized to zero, the regression model becomes computationally cheap and easier to interpret. Thus, LASSO regression is mostly employed in models having large multicollinearity.

8.2.2.2   K-Nearest Neighbors regression

The K-nearest neighbors (KNN) algorithm uses "feature similarity" in order to evaluate the outputs. A new point is evaluated based on the closeness of the points in training data. The mean value of the distances is taken as the final prediction. KNN regression can be explained as a non-parametric method which estimates the relationship between independent variables and the continuous outputs by taking the mean of observations belonging to identical neighborhood [19]. Dimension of neighborhood is set using cross-validation in which the dimension that minimizes mean squared error is selected. Our KNN model utilizes Euclidean distance as the distance metric which is calculated as [20]:

$$\sqrt[2]{\sum_{i=1}^{k} (x_i - y_i)^2}$$

where x is a point of training data, y is a point of test data and k is the number of neighbors. Though effective, the method is unfeasible for data with large dimensions or many independent variables. For a low k-value, the model overfits training data, which increases the error on testing data [21]. However, for a large k-value, the model tends to produce poor results on testing as well as the training dataset. Thus, the error curve reaches a minimum only for a specific k-value, which is 4 in our model (because of 4 input parameters and 4 levels of each input).

8.2.2.3 Support Vector Regression

Support Vector Regression (SVR) shares a similar algorithm as SVMs; besides, it is used to predict discrete values. The main motive of SVR is the evaluation of the best fit line, which is the hyperplane having the maximum number of data points within the boundary line. The closest points on either side of the hyperplane are known as Support Vectors and they help in the determination of the decision boundaries from the input data. Hyperplanes, thus, divide the Euclidean space into separate portions [9,22]. Equation of hyperplane can be represented as:

$$F(w,b) = w^T \varphi(x) + b$$

where w is weight vector, b is bias and $\varphi(x)$ is a non-linear function. SVR aims to find a hyperplane within a threshold distance between hyperplane and boundary line [23]. Kernels are numeric relations which transform the input data into higher dimensional data so as to facilitate the determination of the hyperplane. The kernel is selected based on the problem space and dataset. In our model, the kernel is biquadratic (degree = 4 because of 4 input parameters and 4 levels of each input) with regularization parameter ($C = 2.0$) and error sensitivity parameter ($\varepsilon = 0.1$) set low for wider margins and accurate predictions. The error reduction is achieved by minimizing [22]:

$$\frac{1}{2} w^T \cdot w + C \sum_{i=1}^{l} \xi_i + \xi_i^*$$

The SVR model predicts the values by using the following relation [9,22,23]:

$$f(x) = \sum_{i=1}^{l} \theta_i \varphi(x, x_i) + b$$

8.2.2.4 Artificial Neural Network regression

Artificial Neural Network (ANN) is a set of computational functions which help in the translation of data input into the desired output. Neural networks are made of neurons arranged in multiple layers that have weights assigned at the interlinks. Weight is the learnable parameter that is multiplied by the input received in the neuron. These weights, calibrated with the learning progress, determine the transfer of information from one neuron to another [24]. In our model, there are three layers: the Input layer

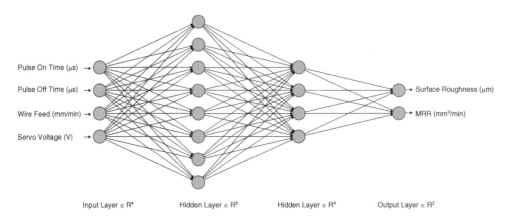

Figure 8.1 Artificial Neural Network structure for modeling WEDM.

consists of 4 neurons (for the four input parameters); the Output layer providing the required outputs consists of 2 neurons (for R_a and MRR); and 2 Hidden layers or the processing layers consist of 8 and 4 neurons, respectively. The aggregation of each layer is passed to the ReLU activation function which then passes the values obtained to the next layer (Figure 8.1).

$$\text{Layer output} = f\left(\sum_{i=1}^{n} w_i x_i + b_i\right)$$

where b is the bias, x is the input to the neuron, w is the weight, n is the number of inputs to this neuron and $f()$ is the activation function [25]. The neural network propagates by adjusting weights to improve the accuracy of the output by minimizing the error which is calculated using the MSE loss function. This modification of weights is achieved using the stochastic optimizer Adam with a learning rate of 0.0003 [26]. The training process is repeated for 3000 epochs, since after that no significant improvement in accuracy or MSE occurs.

8.3   RESULTS AND DISCUSSION

8.3.1   LASSO regression

The MSE values for R_a and MRR produced by LASSO regression are 0.3588 and 0.2407, respectively. Though the error is highly reduced (Figure 8.2), the predicted values do not overlap with the extreme experimental values present in the dataset, especially for surface roughness (Figure 8.3). This occurred because LASSO regression aims to minimize the residual sums of square loss and thus ends up predicting most values near the mean of the dataset. This way, a low error value is produced, but the prediction is not correct for data points in which the values deviate significantly from the mean. For MRR, due to the regular increasing trend, the linear fitting produces significantly better results.

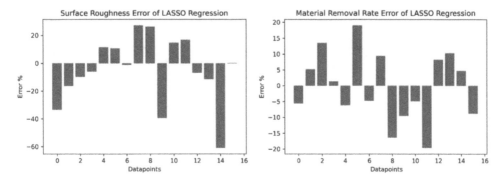

Figure 8.2 Error (%) between actual (experiments) and predicted values (LASSO regression) for R_a (a) and MRR (b).

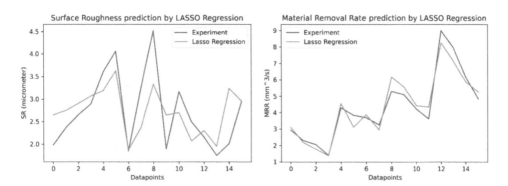

Figure 8.3 Actual (experiments) and predicted values (LASSO regression) for R_a (a) and MRR (b).

8.3.2   K-Nearest Neighbors regression

The MSE values for R_a and MRR produced by KNN regression are 0.4791 and 1.080, respectively. The error values are quite high for most data points (Figure 8.4) and the predicted values do not overlap with experimental values as well (Figure 8.5). Thus, KNN fails to accurately model the WEDM process. This occurred because KNN regression requires large amounts of good quality data. The distribution of all data points very closely deteriorates the results of KNN.

8.3.3   Support Vector Regression

The MSE values for R_a and MRR produced by SVR are 0.3346 and 0.2190, respectively. The error values are high for extreme valued data points only (Figure 8.6). For MRR, the error values are highly minimized and the predicted values also overlap

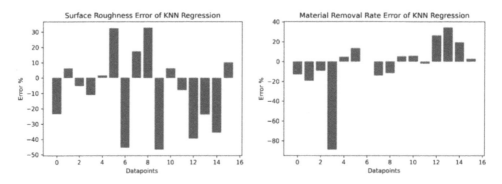

Figure 8.4 Error (%) between actual (experiments) and predicted values (KNN) for R_a (a) and MRR (b).

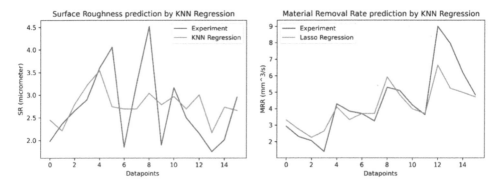

Figure 8.5 Actual (experiments) and predicted values (KNN) for R_a (a) and MRR (b).

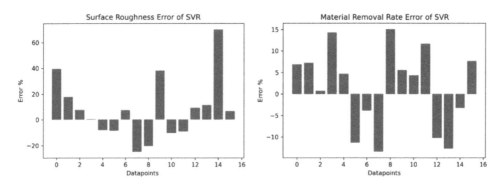

Figure 8.6 Error (%) between actual (experiments) and predicted values (SVR) for R_a (a) and MRR (b).

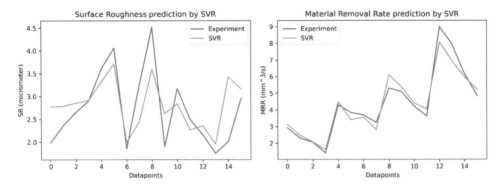

Figure 8.7 Actual (experiments) and predicted values (SVR) for R_a (a) and MRR (b).

with the experimental values; but for surface roughness, the error as well as the overlap is diabolical (Figure 8.7). The surface roughness values do not follow a regular increasing or decreasing trend with any input parameter and are spaced closely with each other, thereby decreasing the kernel efficiency of transforming the data to higher dimensions. The MRR values are evenly spaced, which reduces the occurrence of overlapping. Thus, SVR gives better results for MRR values but doesn't fit very well for surface roughness values.\

8.3.4 Artificial Neural Network regression

The MSE values for R_a and MRR produced by ANN are 0.1425 and 0.2056, respectively. The error values are minimized (Figure 8.8) and the predicted values overlap with the experimental values as well (Figure 8.9). Thus, ANN accurately models the WEDM process, even for such a small dataset. ANN is capable of modeling complex non-linear functions because of different weights and biases associated with the neurons and the activation functions of the layers. After the neural network is trained for

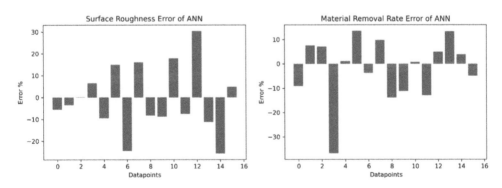

Figure 8.8 Error (%) between actual (experiments) and predicted values (ANN) for R_a (a) and MRR (b).

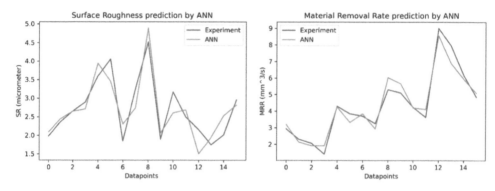

Figure 8.9 Actual (experiments) and predicted values (ANN) for R_a (a) and MRR (b).

a sufficient number of epochs, the optimizer is able to set the weights so as to precisely map any continuous function, which has been proved using the universal approximation theorem [27]. The accuracy of the model will further increase with an increase in the size of the dataset and possible overfitting will also be avoided.

8.4   CONCLUSION

The MSE values for the predicted surface roughness and MRR values, respectively, are (0.36, 0.24) for LASSO regression, (0.48, 1.1) for KNN, (0.33, 0.22) for SVR and (0.14, 0.20) for ANN. Among all the supervised machine learning algorithms utilized in the study, the ANN model predicted the surface roughness as well as MRR with the highest accuracy or overlap and least MSE. Although the training of neural networks requires a greater computation time and power than other algorithms, the difference is insignificant for smaller datasets. Therefore, it can be concluded that artificial neural networks are capable of accurately modeling the WEDM process and effectively replacing the experimental analysis required for the optimum parameter determination. The study, thus, highlights the sustainable impacts of incorporating artificial intelligence with manufacturing techniques, which not only cuts down the time and costs required for the experimentation but also produces parts with high-quality surface finish and dimensional accuracy. Currently, the models can accurately predict the values only if the conditions other than the input parameters to the ML models are kept the same as the training data and the material of the workpiece is H13 die hard steel. With a larger and diversified dataset, the ANN model will be able to predict the output parameters for all combinations of response parameters. If material properties affecting surface roughness and MRR are incorporated into the model, flexibility of workpiece material can also be achieved (Figures 8.10 and 8.11).

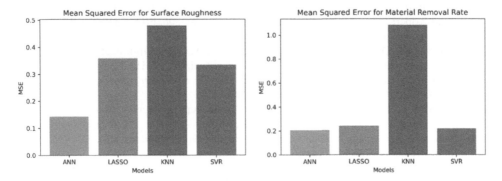

Figure 8.10 MSE between actual (experiments) and predicted values (ML models) for R_a (a) and MRR (b).

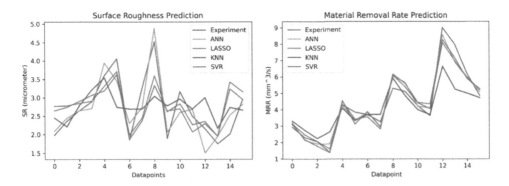

Figure 8.11 Actual (experiments) and predicted values (ML models) for R_a (a) and MRR (b).

REFERENCES

1. Li, S., Wu, X., Chen, S., & Li, J. (2016). Wear resistance of H13 and a new hot-work die steel at high temperature. *Journal of Materials Engineering and Performance*, 25(7), 2993–3006. https://doi.org/10.1007/s11665-016-2124-2.

2. Zhu, J., Zhang, Z., & Xie, J. (2019). Improving strength and ductility of H13 Die Steel by pre-tempering treatment and its mechanism. *Materials Science and Engineering: A*, 752, 101–114. https://doi.org/10.1016/j.msea.2019.02.085.

3. Li, J.-Y., Chen, Y.-L., & Huo, J.-H. (2015). Mechanism of improvement on strength and toughness of H13 Die Steel by Nitrogen. *Materials Science and Engineering: A*, 640, 16–23. https://doi.org/10.1016/j.msea.2015.05.006.

4. Seyedzavvar, M., & Shabgard, M. R. (2012). Influence of tool material on the electrical discharge machining of AISI H13 Tool Steel. *Advanced Materials Research*, 445, 988–993. https://doi.org/10.4028/scientific5/amr.445.988.

5. Shukla, S. K., & Priyadarshini, A. (2019). Application of machine learning techniques for multi objective optimization of response variables in wire cut electro discharge machining operation. *Materials Science Forum*, 969, 800–806. https://doi.org/10.4028/www.scientific.net/msf.969.800.

6. Deshwal, M., Deshwal, D., Kumar, P., & Bhardwaj, V. (2021). Optimization of process parameters for surface roughness and material removal rate of H13 die Tool Steel for Wire EDM using Taguchi Technique. *Journal of Physics: Conference Series*, 1950(1), 012086. https://doi.org/10.1088/1742-6596/1950/1/012086.

7. Kumar, S. (2018). Optimization of wire-EDM process parameters on Machining Die Steel DC53 using Taguchi method with Grey Relational Analysis – A review. *International Journal for Research in Applied Science and Engineering Technology*, 6(6), 427–432. https://doi.org/10.22214/ijraset.2018.6066.

8. Pragnesh, P., Mayank, P., Roshan, P., & Vicky, P. (1970, January 1). Single objective optimization of process parameter of WIRE EDM using response surface methodology. Semantic Scholar.

9. Ulas, M., Aydur, O., Gurgenc, T., & Ozel, C. (2020). Surface roughness prediction of machined aluminum alloy with wire electrical discharge machining by different machine learning algorithms. *Journal of Materials Research and Technology*, 9(6), 12512–12524. https://doi.org/10.1016/j.jmrt.2020.08.098.

10. Singh Nain, S., Sai, R., Sihag, P., Vambol, S., & Vambol, V. (2019). Use of machine learning algorithm for the better prediction of SR peculiarities of WEDM of nimonic-90 superalloy. Archives of Materials Science and Engineering, 1(95), 12–19. https://doi.org/10.5604/01.3001.0013.1422.

11. Phate, M. R., Toney, S. B., & Phate, V. R. (2020). Multi-parametric optimization of WEDM using Artificial Neural Network (ann)-based PCA for Al/SICP MMC. *Journal of the Institution of Engineers (India): Series C*, 102(1), 169–181. https://doi.org/10.1007/s40032-020-00615-1.

12. Gupta, N., Agrawal, A. K., & Walia, R. S. (2019, May). Soft modeling approach in predicting surface roughness, temperature, cutting forces in hard turning process using Artificial Neural Network: An empirical study. In *International Conference on Information, Communication and Computing Technology* (pp. 206–215). Springer, Singapore.

13. Gupta, N., & Walia, R. S. (2021). Predictive soft modeling of turning parameters using artificial neural network. In R. Agrawal et al. (eds.), *Recent Advances in Smart Manufacturing and Materials* (pp. 189–196). Lecture Notes in Mechanical Engineering. Springer, Singapore.

14. Aiyar, H. D. S., Chauhan, G., & Gupta, N. (2021). Soft modeling of WEDM process in prediction of surface roughness using Artificial Neural Networks. In R. Agrawal et al. (eds.), *Recent Advances in Smart Manufacturing and Materials* (pp. 465–474). Lecture Notes in Mechanical Engineering. Springer, Singapore.

15. Melkumova, L. E., & Shatskikh, S. Y. (2017). Comparing Ridge and lasso estimators for data analysis. *Procedia Engineering*, 201, 746–755. https://doi.org/10.1016/j.proeng.2017.09.615.

16. Xin, S. J., & Khalid, K. (2018). Modelling house price using ridge regression and lasso regression. *International Journal of Engineering & Technology*, 7(4.30), 498. https://doi.org/10.14419/ijet.v7i4.30.22378.

17. Muthukrishnan, R., & Rohini, R. (2016). Lasso: A feature selection technique in predictive modeling for Machine Learning. *2016 IEEE International Conference on Advances in Computer Applications (ICACA)*. https://doi.org/10.1109/icaca.2016.7887916.

18. Mangalathu, S., Jeon, J.-S., & DesRoches, R. (2017). Critical uncertainty parameters influencing seismic performance of bridges using Lasso regression. *Earthquake Engineering & Structural Dynamics*, 47(3), 784–801. https://doi.org/10.1002/eqe.2991.

19. Goyal, R., Chandra, P., & Singh, Y. (2014). Suitability of KNN regression in the development of interaction-based software fault prediction models. *IERI Procedia*, 6, 15–21. https://doi.org/10.1016/j.ieri.2014.03.004.

20. Song, J., Zhao, J., Dong, F., Zhao, J., Qian, Z., & Zhang, Q. (2018). A novel regression modeling method for PMSLM structural design optimization using a distance-weighted KNN algorithm. *IEEE Transactions on Industry Applications*, 54(5), 4198–4206. https://doi.org/10.1109/tia.2018.2836953.

21. Wang, P., Zhang, Y., & Jiang, W. (2021). Application of K-nearest neighbor (KNN) algorithm for Human Action Recognition. *2021 IEEE 4th Advanced Information Management, Communicates, Electronic and Automation Control Conference (IMCEC)*. https://doi.org/10.1109/imcec51613.2021.9482165.

22. Bilhan, O., Emiroglu, M. E., Miller, C. J., & Ulas, M. (2018). The evaluation of the effect of nappe breakers on the discharge capacity of trapezoidal labyrinth weirs by Elm and SVR approaches. *Flow Measurement and Instrumentation*, 64, 71–82. https://doi.org/10.1016/j.flowmeasinst.2018.10.009.

23. Baydaroğlu, Ö., & Koçak, K. (2014). SVR-based prediction of evaporation combined with chaotic approach. *Journal of Hydrology*, 508, 356–363. https://doi.org/10.1016/j.jhydrol.2013.11.008.

24. Farizawani, A. G., Puteh, M., Marina, Y., & Rivaie, A. (2020). A review of artificial neural network learning rule based on multiple variant of conjugate gradient approaches. *Journal of Physics: Conference Series*, 1529(2), 022040. https://doi.org/10.1088/1742-6596/1529/2/022040.

25. Moolayil, J. (2018). Deep Neural Networks for supervised learning: Regression. *Learn Keras for Deep Neural Networks*, 53–99. https://doi.org/10.1007/978-1-4842-4240-7_3.

26. Kingma, D. P., & Ba, J. (2015). Adam: A method for stochastic optimization. In: Y. Bengio, Y. LeCun (eds.) *3rd International Conference on Learning Representations (ICLR)*, May 7–9, 2015, San Diego, CA.

27. Kratsios, A. (2021). The universal approximation property. *Annals of Mathematics and Artificial Intelligence*, 89(5–6), 435–469. https://doi.org/10.1007/s10472-020-09723-1.

Chapter 9

Investigation of dragline productivity

*Prashant Washimkar, Pramod Belkhode,
Ayaz Afsar, and Kanchan Borkar*

CONTENTS

9.1   INTRODUCTION

The productivity is not satisfactory as compared to the use and cost of the available dragline [1–4]. Investigation of this phenomenon is necessary to achieve maximum productivity and to improve the present method of maintenance operation to reduce the downtime failure by identifying the failure [5,6]. The effectiveness and utility of the maintenance schedule at all levels are also estimated. The most affecting component is identified to enhance productivity and to minimize the overhauling time so that the human energy can be minimized.

Equipment breakdown (or shutdown) list for a period of approximately one and a half years has been analyzed for the two important subsystems, namely Drag and Hoist. These analyses revealed two significant failures, i.e., drag rope and oil ring for Drag and spair link for Hoist. Presently, routine maintenance action is performed on the various components of drag and hoist without prioritizations [7,8]. Hence, suitable preventive maintenance action along with the schedule is established on a priority basis. Monthly check-up of rope, oil ring and spair link will take 2 hours/month as preventive maintenance downtime [9–12]. Hence, preventive maintenance schedule design results in saving potential of Rs. 1.8 crores per annum.

The maintenance schedule is a design, which is based on the activities of the dragline that help to improve the performance of the various components involved in

DOI: 10.1201/9781003293576-9

the dragline. This novel method of schedule maintenance improves the performance analysis, reduces the failure time and enhances productivity. The analysis is done on the recorded observations of one and a half years in which the main cause of the breakdown is the failure of the dragline component due to improper maintenance schedule resulting in the poor productivity. The breakdown time is reduced by reducing the failure time from 88 hours to 36 hours, particularly for these three components – drag rope, oil ring and spair link – which further reduced the potential cost from 4.4 crores to 1.8 crores due to a reduction in the breakdown time. This novel concept of maintenance is effective to reduce the downtime and enhance productivity meaning the overall efficiency of the workstation is improved.

9.2   DRAGLINE

Dragline system weighs about 3,000 tonnes and has a boom length between 87 and 99 m. The maximum allowable load that is recommended by the Original Equipment Manufacturer (OEM) is in the range of 115.7–138 tonnes. In open cast coal mining, the dirt (or overburden) above the coal seam needs to be removed so that the coal can be reached. The mining process occurs in strips and is often referred to as strip mining [13–18]. The overburden is blasted to break it up along large strips. The blasted overburden is reformed as muck, and the pile of overburden is referred to as a muck pile. The operation of the dragline involves four different types of motion – dragline, hoist, swing and propel. The operation of the dragline is cyclic in nature. The dragline is moved (propelled) to the position from which it will dig. The dragline then removes the muck from above the coal seam and dumps it in the previous strip where the coal has been removed [19–22]. This cycle of digging and dumping of the muck is called the duty (dig) cycle.

The duty cycle can be divided into the following phases:

1. **Bucket spotting:** The bucket is correctly aligned with the muck to be excavated in the next dig cycle.
2. **Bucket filling:** The bucket is dragged through the muck pile until full. The bucket is raised (hoisted) to a specific height before the dragline begins to swing.
3. **Swing to dump:** The dragline then swings around to the dumping position.
4. **Dump:** The bucket is tilted and the muck dumped.
5. **Return swing:** The dragline swings around to the next digging position.

It is important to note that the bucket filling time, the hoisting time and swing to dump time are all dependent on the dragline load. After the dragline has moved all the overburden it can from its position, it moves along the bench in the direction it is placing the muck so that the dumped muck becomes the new pad on which the dragline will rest [23,24]. The dragline moves along until the mining of the strip is complete and then the mining cycle repeats (Figures 9.1 and 9.2).

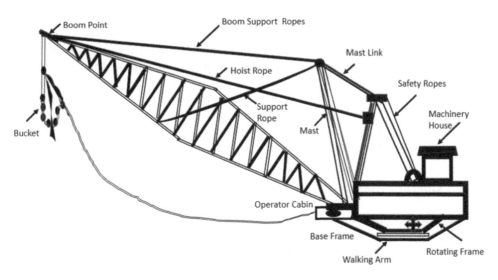

Figure 9.1 Dragline.

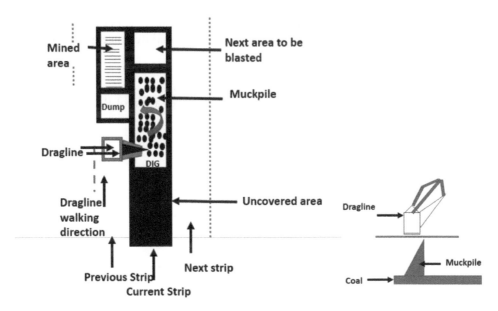

Figure 9.2 Muck pile.

9.3 METHODOLOGY

The research of dragline operations involved the observation of breakdown failure which results in the design of the maintenance schedule for optimization of the failure time [25,26]. The analysis of the observed data for monitoring the dragline, particularly breakdown time and cost analysis, is associated with optimizing the failure cost. Breakdown time for subsystems drag and hoist is much more; hence, they are selected for estimation of reliability and its improvement in the development of preventive maintenance schedule [27,28]. Mean Time Between Failures (MTBF) for the prominent failure modes is estimated as shown in Tables 9.1 and 9.2.

9.3.1 Current maintenance scenario

Maintenance action is in the form of routine maintenance as per described schedule. The existing maintenance schedule is as follows:

1. Greasing of pulleys over which drag or hoist rope moves
2. Greasing of roller
3. Checking of bucket for any damage
4. Checking of drag rope and hoist rope for torn condition
5. Checking for oil leakage
6. Cleaning of machine

Equipment shutdown list for a period of about one and a half years was obtained and the data regarding various breakdowns and associated maintenance action was analyzed as shown in Table 9.3.

Total breakdown time = 126.5 hours
If the dragline system remains file for 1 hour, it causes a loss of Rs. 5 lacs.

Therefore, present loss $= 126.5 \times 5$
$$= Rs. \ 632.5 \ lacs = Rs. \ 6.325 \ crores$$

Table 9.1 Drag and hoist

S. no.	Subsystem	Failure item	MTBF (monthly)
01	Drag	Drag rope	1
		Oil ring	5
02	Hoist	Spair link	2

Table 9.2 Drag and hoist frequency failure

S. no.	Subsystem	Failure item	Frequency of maintenance failure
01	Drag	Drag rope	4
		Oil ring	4
02	Hoist	Spair link	6

Table 9.3 Equipment failure time

S. no.	Subsystem	Item failed	Failure time in hours (idle time)
01	**Drag**	Drag brake air pipe leakage	I hour
		Drag not working properly (Electrical Problem)	2 ½ hours
		Drag reduction gearbox shaft broken	28 ½ hours
		Drag rope fitted	3 ½ hours
		Drag rope fitted	14 hours
		Drag rope fitted	24 hours
		Drag spair link fitted	I hour
		Drag rope fitted	5 hours
		Drag rope out from hitch	2 hours
		Drag O ring broken and fitted	3 hours
		Drag O ring fitted	2 hours
		Drag new rope fitted	6 ½ hours
		Drag spair link broken	3 hours
		Drag chain not working	2 hours
		Drag spair link broken	2 hours
		Drag O ring pin fitted	2 hours
		Drag brake problem	I hour
		Drag O ring broken	I ½ hours
		Drag O ring broken	4 hours
		Drag O ring tooth point fitted	2 hours
		Drag coupling pad broken	2 hours
02	**Hoist**	Hoist circuit earth checking	½ hour
		Hoist spair link fitted	I hour
		Hoist spair link broken and fitted	3 hours
		Hoist chain link pin fitted	I hour
		Hoist spair link broken	I hour
		Hoist spair link fitted	I hour
		Hoist spair link fitted	I hour
		Hoist spair link fitted	I hour
		Hoist spair link broken	I hour
		Hoist spair link pin fitted	I hour
		Hoist spair link fitted	2 ½ hours
Total breakdown time of drag and hoist			**126.5 hours**

9.3.2   Electric supply

The dragline takes power directly from the power grid through large electric cables [29]. The AC power is converted to the DC power through the use of Motor-Generator (MG) sets. Within each MG set are generators for the drag, hoist and swing motions. The MG set generators are connected to DC motors, which are connected to appropriate motion gears. The propel gears are connected to the "feet" used to propel the dragline and the swing plenary gears rotate the dragline [30,31]. The drag and hoist gears are connected to drums on which the drag and hoist ropes are wound. The drag and hoist ropes are connected to the bucket and control its position.

9.4   ANALYSIS OF DRAGLINE

In the entire dragline system, drag and hoist are selected for further improvement potential as breakdown time with these two subsystems is much more [32,33]. It is very important to reduce the breakdown time of drag and hoist.

9.4.1   Analysis of the data

The analysis of the data revealed two significant failures – drag rope and oil ring. For drag (Figures 9.3 and 9.4) and spair link of hoist (Figures 9.5 and 9.6).

It is clarified from the graph that the frequency of failure for items, namely drag rope, oil ring and spair link is much more. Idle time for the items is as follows.

Idle time for drag rope = 63 hours
Idle time for oil ring = 14.5 hours
Idle time for spair link = 10 hours
Therefore, total idle time = 87.5 ≈ 88 hours.

∴ Total revenue loss = 88 hours × 5 lacs / hour
= Rs. 440 lacs = Rs. 4.4 crores

Hence, it is necessary to minimize the revenue losses to increase saving potential. The proposed preventive maintenance schedule requires downtime of approximately 2 Rs/month; hence, the total downtime for the period of one and a half years is 36 hours. The downtime cost will be Rs. 1.8 crores (since downtime cost per hour is Rs. 5 lacs).

9.4.2   Calculation of repair cost

$$\text{Average time of repair} = \frac{\text{Total breakdown / repair time}}{\text{Frequency of failure}}$$

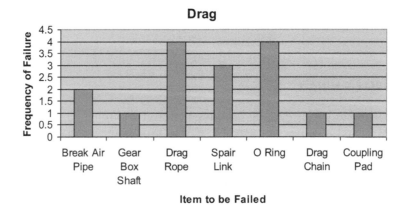

Figure 9.3 Graph of item to be failed vs frequency of failure for drag.

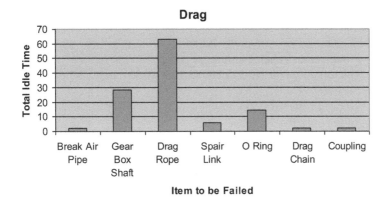

Figure 9.4 Graph of item to be failed vs total idle time for hoist.

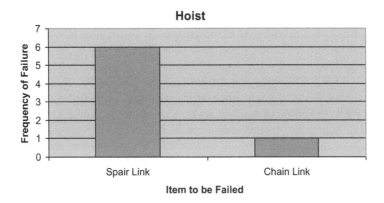

Figure 9.5 Graph of item to be failed vs frequency of failure for drag.

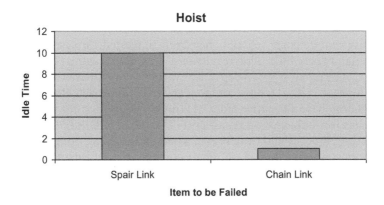

Figure 9.6 Graph of item to be failed vs total idle time for hoist.

1. Average financial individual repair cost
    = (Average time of repairs of Drag rope+Spare link+O ring)×Downtime cost/hour
    $= (2.83+0.9+2.357) \times 100,000 = \text{Rs. } 608,700$
2. Average group repair/replacement cost
    $= (2.83 \times 100,000)+((0.11 \times 214,500)/3.42)+((0.32 \times 16,000)/3.63)$
    $\approx \text{Rs. } 291,309.59$
    Cost of drag rope=Rs. 214,500
    Cost of O ring=Rs. 16,000
3. Total amount saved=Rs. 608,700−Rs. 291,309
    = Rs. 317,390 (for 4 months)
    $\approx$ Rs. 9.5 lacs/year (approx.)

By carrying out cost analysis, net financial saving of Rs. 9.5 lacs/year (approx.) is estimated.

9.5   RESULT AND DISCUSSION

The revenue losses are minimized to increase saving potential in which the proposed preventive maintenance schedule requires downtime of approximately 2 Rs/month; hence, the total downtime for the period of one and a half years is 36 hours. The downtime cost will be Rs. 1.8 crores (downtime cost per hour is Rs. 5 lacs). Presently, the total downtime for the failure is 88 hours for three items (namely drag rope, oil ring and spair link) which cost a loss of Rs. 4.4 crores. The saving potential is Rs. 2.6 crores (for 18 months), i.e., nearly Rs. 1.8 crores/annum.

The result of the present study is compared with the current research in terms of breakdown time and costs associated with the different draglines located in the nearby region and data is verified with the current research result on a dragline. The specific drag and hoist are being studied with respect to the failure time and loss associated with it during the breakdown time. Based on that, cost analysis is done which proves that a net financial saving of Rs. 9.5 lacs/year (approx.) is estimated.

9.6   CONCLUSION

The preventive maintenance schedule has been suggested for systems, namely drag rope, oil ring and spair link. This results in a maximum saving potential of Rs. 1.8 crores/annum. The repair cost is a dependent variable, and failure time is an independent variable for each critical failure component. In this approach, drag rope, spare link and O ring of drag have a negative index of failure time. Hence, for these components, failure time has less influence on repair cost. The index of failure time is positive for components drag bracket and drag brake, which indicates a strong impact on repair cost. It represents a lesser impact on repair cost for components namely drag brake. Similarly, for hoist spare link, a negative index of failure time indicates less influence on the repair cost.

REFERENCES

1. Samanta, B., & Sarkar, B. (2002). Availability modeling of a drag line system – case study. *Journal of the Institution of Engineers (India)*, 83(1), 20–26.
2. Belkhode, P., Sakhale, C., & Bejalwar, A. (2020). Evaluation of the experimental data to determine the performance of a solar chimney power plant. *Materials Today: Proceedings*, 27, 102–106. doi:10.1016/j.matpr.2019.09.006.
3. Dhutekar, P., Mehta, G., Modak, J., Shelare, S., & Belkhode, P. (2021). Establishment of mathematical model for minimization of human energy in a plastic moulding operation. *Materials Today: Proceedings*. doi:10.1016/j.matpr.2021.05.330.
4. Belkhode, P. N., Ganvir, V. N., Shende, A. C., & Shelare, S. D. (2022). Utilization of waste transformer oil as a fuel in diesel engine. *Materials Today: Proceedings*, 49, 262–268. doi:10.1016/j.matpr.2021.02.008.
5. Belkhode, P. N. (2019). Optimum choice of the front suspension of an automobile. *Journal of Engineering Sciences*, 6(1), e21–e24. doi:10.21272/jes.2019.6(1).e4.
6. Mathew, J. J., Sakhale, C. N., & Shelare, S. D. (2020). Latest trends in sheet metal components and its processes—A literature review. *Algorithms for Intelligent Systems*, 565–574. doi:10.1007/978-981-15-0222-4_54.
7. Modak, J. P., Mehta, G. D., & Belkhode, P. N. (2004). Computer aided dynamic analysis of the drive of a chain conveyor. *Manufacturing Engineering and Materials Handling Engineering*. doi:10.1115/imece2004-59157.
8. Waghmare, S., Mungle, N., Tembhurkar, C., Shelare, S., Sirsat, P., & Pathare, N. (2019). Design and analysis of power screw for manhole cover lifter. *International Journal of Recent Technology and Engineering*, 8(2), 2782–2786, doi:10.35940/ijrte.B2628.078219.
9. Modak, J. P. Project Proposal on Condition Monitoring of Drag Line of WCL; Nagpur, 2005.
10. Sakhale, C. N., Sonde, V., & Belkhode, P. N. (2015). Physical and mechanical characteristics for cotton and pigeon pie as agriculture residues. *International Journal of Application or Innovation in Engineering & Management (IJAIEM)*, 4(7), 156–169.
11. Sahu, P., Shelare, S., & Sakhale, C. (2020) Smart cities waste management and disposal system by smart system: A review. *International Journal of Scientific & Technology Research*, 9(03), 4467–4470.
12. Townson, Load – Maintenance Interaction: Modelling and Optimisation; A thesis submitted to the university of Queensland for the degree of Doctor of Philosophy, 2002.
13. Belkhode, P. N. (2017). Mathematical modelling of liner piston maintenance activity using field data to minimize overhauling time and human energy consumption. *Journal of the Institution of Engineers (India): Series C*, 99(6), 701–709. doi:10.1007/s40032-017-0377-7.
14. Modak, J. P., Belkhode, P. N., Bodhankar, D., Himte, R. L., & Washimkar, P. V. (2008). Modeling and analysis of front suspension for improving vehicle ride and handling. *2008 First International Conference on Emerging Trends in Engineering and Technology*. doi:10.1109/icetet.2008.136.
15. Waghmare, S., Shelare, S., Sirsat, P., Pathare, N., & Awatade, S. (2020). Development of an innovative multi-operational furnace. *International Journal of Scientific & Technology Research*, 9(4), 885–889.
16. Jawalekar, S. B., & Shelare, S. D. (2020). Development and performance analysis of low cost combined harvester for Rabi crops. *Agricultural Engineering International: CIGR Journal*, 22(1), 197–201.
17. Belkhode, P.N. & Modak, J.P. (2011). Comparison of steering geometry parameter of front suspension of automobile. *Internal Journal of Scientific and Engineering Research*, 3, 1–3.

18. Naidu, V. A. M., Shelare, S. D., & Awatade, S. (2020). A review on design of components of 4 stroke engine using hybrid metal matrix. *International Journal of Mechanical and Production Engineering Research and Development*, 10(3), 8853–8862. doi:10.24247/ijmperdjun2020842.

19. Shelare, S. D., Aglawe, K. R., Waghmare, S. N., & Belkhode, P. N. (2021). Advances in water sample collections with a drone – A review. *Materials Today: Proceedings*, 47, 4490–4494. doi:10.1016/j.matpr.2021.05.327.

20. Belkhode, P., Ganvir, V., Shelare, S., Shende, A., & Maheshwary, P. (2022). Experimental investigation on treated transformer oil (TTO) and its diesel blends in the diesel engine. *Energy Harvesting and Systems*, 9(1), 1–11. doi:10.1515/ehs-2021-0032.

21. Belkhode, P. N. (2019). Development of mathematical model and artificial neural network simulation to predict the performance of manual loading operation of underground mines. *Journal of Materials Research and Technology*, 8(2), 2309–2315. doi:10.1016/j.jmrt.2019.04.015.

22. Vidyasagar, V., Belkhode, D. P. N., & Modak, D. J. P. (2014). Mathematical model for manual loading activity in underground mines. *International Journal of Mechanical and Industrial Engineering*, 134–141. doi:10.47893/ijmie.2014.1199.

23. Belkhode, P. N. (2019). Analysis and interpretation of steering geometry of automobile using artificial neural network simulation. *Engineering*, 11(04), 231–239. doi:10.4236/eng.2019.114016.

24. Shelare, S. D., Kumar, R., & Khope P. B. (2021). Assessment of physical, frictional and aerodynamic properties of charoli (Buchanania Lanzan Spreng) nut as potentials for development of processing machines. *Carpathian Journal of Food Science and Technology*, 174–191. doi:10.34302/crpjfst/2021.13.2.16.

25. Belkhode, P. N., Shelare, S. D., Sakhale, C. N., Kumar, R., Shanmugan, S., Soudagar, M. E. M., & Mujtaba, M. A. (2021). Performance analysis of roof collector used in the solar updraft tower. *Sustainable Energy Technologies and Assessments*, 48, 101619. doi:10.1016/j.seta.2021.101619.

26. Shelare, S. D., Aglawe, K. R., & Belkhode, P. N. (2021). A review on twisted tape inserts for enhancing the heat transfer. *Materials Today: Proceedings*. doi:10.1016/j.matpr.2021.09.012.

27. Shelare, S. D., Aglawe, K. R., & Khope, P. B. (2021). Computer aided modeling and finite element analysis of 3-D printed drone. *Materials Today: Proceedings*, 47, 3375–3379. doi:10.1016/j.matpr.2021.07.162.

28. Khope, P. B., & Shelare, S. D. (2021). Prediction of torque and cutting speed of pedal operated chopper for silage making. *Advances in Industrial Machines and Mechanisms*, 89–97. doi:10.1007/978-981-16-1769-0_22.

29. Nadarajah, S., Ali, M., & Kotz, S.. (2000). On the ratio of Weibull random variables with dragline application. *International Journal of Industry Engineering*, 14(1), 84–89.

30. Oke, S. A., & Charles-Owaba, O. E. (2006). An approach for evaluating preventive maintenance scheduling cost. *International Journal of Quality & Reliability Management*, 23(7).

31. Shankar, G., & Sahani, V. (2003). Reliability analysis of a maintenance network with repair and preventive maintenance. *International Journal of Quality & Reliability Management*, 20(2).

32. Waghmare, S. N., Sakhale, C. N., Tembhurkar, C. K., & Shelare, S. D. (2019). Assessment of average resistive torque for human-powered stirrup making process. *Computing in Engineering and Technology*, 845–853. doi:10.1007/978-981-32-9515-5_79.

33. Belkhode, P., Modak, J. P., Vidyasagar, V., & Shelare, S. (2021). Procedure of collecting field data: Causes, extraneous variables, and effects. *Mathematical Modeling and Simulation*, 33–47. doi:10.1201/9781003132127-5.

Chapter 10

Lean administration in the Order-to-Cash process

Soumil Mukherjee, Vikram Sharma, and Ravi Prakash Gorthi

CONTENTS

10.1 INTRODUCTION

Given the amount of research and articles on lean management, the significance of lean techniques in office and work processes to reduce waste and enhance productivity is widely established. Majorly applied in the manufacturing industries, lean practices in waste reduction ensure that the remaining steps flow along with continuous improvement in the production system. The term 'waste' here refers to the activities that neither create value for the customer nor are required in the work process; instead, it slows down the entire process. For instance, mistakes that need rectification, employees' waiting time in the downstream activities due to delays in the upstream activities, etc. [1]

Waste is an unavoidable part of any manufacturing sector's O2C process, including the pharmaceutical component manufacturing industries, given its complexities in managing manual activities across the variegated steps and systems departments. The O2C process refers to a firm's business process for the entire order processing system. It starts with the inquiry of a particular product, quotation, order placement by the customer, preparing for shipment, dispatch of the order, and final invoicing. To effectively fulfil the O2C process, multiple steps and departments are involved, such as sales, service, finance, engineering, operations, and logistics. The process mainly integrates the sales and finance departments, involving several organizational levels in between, namely, company code, sales organizations, distribution channels, divisions, and plants [2].

Nonetheless, given the diversity of departments, steps, and levels associated with the O2C process, there are bound to be specific issues leading to inefficiencies through increased waste. Some of these issues are as follows: a lack of standard end-to-end O2C processes, a wide variety of enterprise resource planning systems in use,

DOI: 10.1201/9781003293576-10

issues in delivery fulfilment, a high number of errors while taking orders, issues in master data, high volume of manual entries, higher days sales outstanding, lack of clearly defined and meaningful key performance indicators, and missing of a robust control system [3].

In the manufacturing sector, lean practices act in the systemic elimination of waste from operations by maintaining a synergy between the production of products or services and the demand rate. Nonetheless, given the broad arena of the manufacturing sector, with a host of industries coming within its fold, implementation of the lean practices is carried out keeping variegated purposes in mind, with the core purpose being streamlining operations. For instance, the mediating role of lean manufacturing between IT investments in advanced manufacturing technology and the improvement of business performance [4]. Past research on implementing lean management has focused on drug-producing companies concerning the pharmaceutical industry. For instance, the case study presented by Nenni, Giustiniano, and Pirolo [5] focuses on manufacturing pharmaceutical products like capsules, pills, tablets, etc., and unearths the benefits of implementing lean techniques in gaining efficiency improvement and competitive advantage.

The O2C process in the manufacturing sector involves waste owing to the complex interaction of the in-built processes of information management, data management, inter-departmental involvement in processing orders to final delivery of the products, and a host of other such functions. Regardless of the fact, studies on the implementation of lean management system in the O2C process or its impact is relatively low. There is a plethora of research on the O2C process, considering aspects [3,6] and the role of lean information management in handling information waste in generic manufacturing enterprises. This occurs due to broken information management, lack of understanding of which information should flow, overproduction or overflow of information, and inaccurate flow. The study showcases the significance of lean information management in identifying waste and the root cause generating such waste, thereby helping the organizations save resources (like rework, inspection, and correction of information), time, and ensure efficient information quality and effective communication among departments and stakeholders.

When understanding the role of lean management in the O2C process of the pharmaceutical component manufacturing industry, there is no such evidence present, hence making this study the first to research the concerned issue. Even from the pharmaceutical industry's perspective concerning the O2C process, Moosivand, Ghatari, and Rasekh [7] bring out the supply chain-related challenges like inaccuracy in forecasting, long lead times, lack of optimum target inventory, and high supply chain costs. However, the study does not deal with the implementation of lean administration in addressing the issues highlighted. The review lacked in developing background on the role of lean practices in addressing issues like parallelism in work areas among stakeholders of O2C, manual record-keeping, considerable paperwork in logistics, management of heavy email traffic, double effort in data entry, lack of clarity in Key Performance Indicators (KPIs), robust control mechanisms and lack of Document Management System (DMS) and Customer Relationship Management system (CRM). Broadly, the studies above focus on manufacturing pharmaceutical drugs and not on components [8,5,9,10]. Also, none of the studies on lean practices above focuses on the O2C process optimization using Swimlane mapping and Kaizen bursts to identify issues at the initial stage of implementation.

Observations from the review of existing literature show that the pharmaceutical component manufacturing industry is still an isolated area where limited studies have been carried out to establish the implementation of lean practices and their impact. Given the considerable presence of the pharmaceutical component manufacturing industry in the global business platform and its contribution to bringing out novel technologies, machines, and similar components thereby assisting the healthcare sector in its efficient functioning [10], research on the impact of lean practices in streamlining the industry's operations upon understanding its core functional issues is imperative. The O2C process is a core component of product flow within the manufacturing sector. Research on the current form of wasteful activities and their elimination through lean implementation will provide new insights to the industry stakeholders to identify waste and improve existing processes.

10.2 METHODOLOGY

The methodology of this case study comprised of three phases—identification of issues in the O2C process, then, analysis of the issues by breaking them into individual activities carried out daily by the concerned employees and efforts they invested concerning Full Time Equivalents (FTEs) and, finally, observation of the efficiency of the lean-based future-state strategies in addressing the operational issues.

The first phase comprised of the 5-day workshop conducted by Company X as part of its plan to implement lean production. This workshop acted as a tool to identify the present condition (as-is) and improvement areas (to-be) to identify the waste and value-added and non-value-added activities, thereby transforming them into value-added ones. As part of the workshop, observations and interactions were made with the O2C stakeholders of Company X to understand the work process across the three departments of sales, logistics, and finance.

Both the current issues and the improvement areas in the O2C processes were mapped using the swim lane lean management tool. The mapping process in swim lane facilitates plotting and tracing activities quickly. Most importantly, this mapping process projects the interconnections between departments, teams, and their activities. Through a process flowchart, swim lane provides enriched information on which unit or team carries out which function, as per their sequence [11]. Each of the activities within and between departments is present within a single swim lane. The interaction is shown through dashed lines connecting corresponding steps in different swim lanes, referred to as 'cross entity' messages (Figure 10.1).

In the first process of identification of issues, the As-Is Swim Lane map was developed through data analysis carried out of observations on 687 customer orders with 2.376 order lines operated. With nearly 3.4 order lines per Purchase Order (PO), there were about 10 order lines per day on average. Further, out of 63 articles (products and materials) and 106 customers, 80% of the net value was made with the top 12 products and 18 customers.

This swim lane map also involved several Kaizen bursts, as planned through the workshop. These Kaizen bursts acted as part improvement steps (which can be addressed through individual mini-workshops) that led to the achievement of the overall/final target. Kaizen clouds or bursts are based on the philosophy of Kaizen learning which refers to continuous improvement [12], which is based on the reflective learning cycle grounded on quality assurance methodologies.

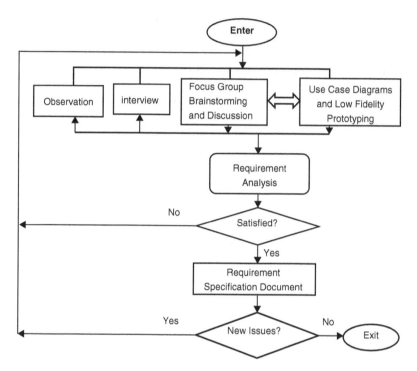

Figure 10.1 Process diagram to develop swim lane activity map.

As part of its plan to implement lean production techniques, Company X developed specific effectiveness teams called 'Kaizen teams' during this 'Kaizen workshop'. Through the application of Kaizen bursts, these 'Kaizen teams' identified nearly 50 mini-improvement areas in the fields of IT automation, data quality, IT tools, paperwork, diverse challenges, transparency, process, knowledge waiting/movement, and standards. This identification of the mini-improvement areas through Kaizen bursts later guided the development of the future-state (to-be) swim lane, implementing the lean technique and reducing the waste activities, thereby creating time-based and economic-financial savings.

In the second phase, an ABA was conducted with 33 internal stakeholders involved in the O2C process across the three departments of Company X. The ABA refers to 'treating individual actions in space and time' to understand human mobility and communication behaviour along with the related systems like the physical environment [3]. The approach enables forecasting and knowledge construction by capturing complex issues as constraints and focusing on individual participation in limited availability in time and space. The analysis here purported to identify the effort per FTE invested by the employees in carrying out their daily activities so that waste activities can be separated from the value-added ones. Through a questionnaire sent via emails, the participants were asked to record and categorize their daily efforts for 2 weeks into value-added, non-value-added/parallel, and wasteful activities. Analysing the categorized activities received from the 32 responses (97% response rate), allocating 31.5 FTEs identified the actual waste and value-added activities and time invested in carrying out the same. The ABA enabled the researcher to develop strategies to eliminate waste activities and utilize the effort to create value-added activities.

The third phase consisted of developing the future-state process of lean administration based on the mini-improvement areas identified by the Kaizen bursts and findings of the ABA. The results of the lean administration were later developed into the future-state (to-be) swim lane to showcase comparatively if the implementation of the lean technique was beneficial to the company.

The entire study involving the three stages of observation took place in 12 months with the first and second stages taking place within 10 days and the third stage after 1 year, to provide the organization scope of improvement on the Kaizen bursts identified through the workshop and future-state strategies developed thereafter. During these 12 months of post-workshop and future-state strategy development, regular follow-ups were conducted with the concerned group for participants to measure the developments in terms of operational excellence in the O2C process taking place.

10.3 IMPLEMENTATION

10.3.1 Phase I: understanding the As-Is scenario through swim lane and Kaizen bursts

The As-Is Swim Lane map highlighted the present activities across sales, logistics, and finance departments in Company X, engaged in the O2C processes of inquiry, order entry, dispatch, and invoice clearing. The entire swim lane map depicted overlapping activities among departments, with each person handling different tasks, logistics involving too much paperwork, heavy traffic of emails, and one of the leading systems of carrying out the O2C processes (Figure 10.2).

The Kaizen bursts brought 50 mini-improvement areas (wasteful activities). Company X dealt with a large amount of manual processing and paperwork in the form of financial documents, manual documents in logistics, picking lists, printouts of each PO, and a series of paperwork in the quality department. The quality of data was

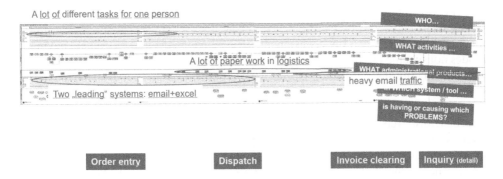

Figure 10.2 Swim Lane As-Is.

marred by the unnecessary effort being made to cross-verify prices against pricing index data maintained in the company database.

A robust customer relationship management system will enable the organization to manage customer interactions by focusing on transactional exchanges. Further integrating the DMS with the CRM processes enable the organization to evaluate core business measures such as customer profitability, satisfaction, and loyalty to sustain its business decisions. Managing customer interactions through CRM processes in information sharing, customer involvement, long-term partnership, joint problem-solving, and technology-based CRM positively impacts organizational innovation [1]. This innovation is observed in the organization's product, process, administrative, marketing, and service capabilities.

Kaizen bursts on IT tools showed that a PowerPoint file used to take nearly 1.5 hours to be prepared with the involvement of four participants. In terms of knowledge, not every concerned employee of the three departments involved in the O2C process understands the incoterms. For instance, the sales personnel were ignorant of the incoterms' in-depth understanding. In certain situations, incoterms and payment terms did not match, namely cash against documents, ex-works, etc. Some important documents were not made available to everybody involved in the O2C process but were stored locally. An adjustment in credit limit was carried out only by some specific personnel. All these cavities reflect some severe problems in the knowledge management processes in Company X, which further contributes to the non-maintenance of standardized prices, incoterms—leading to their irregular usage, and freight cost, e.g., the addition of up to 20% in the freight cost most recently. Studies show the positive influence of a systematic approach in combining the ideals of knowledge management and cross-functional teams (here, the four departmental teams involved in the O2C process) on the organization's performance [13]. Applying an efficient knowledge management system enables the organization to enhance the flow of knowledge, thereby enhancing the cost, time, and quality involved in the production, which in turn cumulatively contributes to the competitive advantage and hence performance [14]. Therefore, the requirement of an efficient knowledge management system within the O2C process of Company X is evident to ensure that the organization's overall competitive advantage and performance are not hampered.

Besides these, several issues were related to the waiting time or product movement after the product got dispatched. It took nearly 2 days to get the freight cost back from logistics, and several follow-ups and constant reminders were required to obtain shipping documents. Usage of a single printer among the departments involved in the O2C process caused further order delays. Then, diverse payment-related issues in the form of difficulties when the government released payments and customers paying in modes not specified in the invoices. These issues affected the overall O2C process from efficient functioning due to non-clarity in communication both inter-department and with the customers.

A significant improvement area reflected by the Kaizen bursts was the condition of transparency within the organization. Transparency regarding the strategic goals, value chain, associated end-to-end processes, and KPIs is significant to ensure an efficient O2C process. However, Company X was facing process transparency-related issues in the form of non-clarity of personnel working on and having access to the company's overdue report; no database on exact product quotation; and lack of

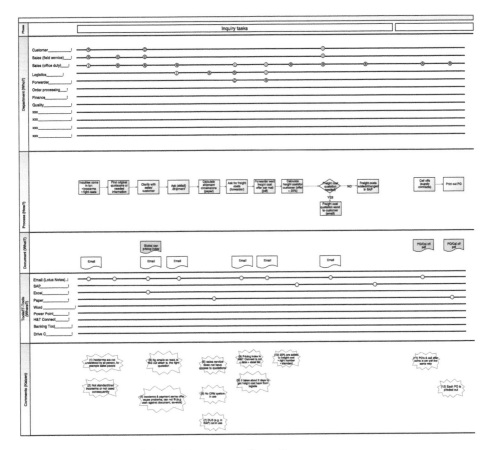

Figure 10.3 Evaluation of the As-Is process (Part I).

communication on open quality issues or product delivery status. Unavailability of an exact quotation database additionally led to their inaccessibility to the sales service. In the case where the PO was unclear, quotation was collected with the help of sales field employees. Overdue was another result of process non-transparency. Non-collected ex-works (EXW) deliveries caused a financial burden on the organization in case service fee was not charged from the buyer (Figures 10.3 and 10.4).

10.3.2 Phase II: identification of activities/process steps as VA, NVA, and W

Through the ABA, the daily efforts of the concerned employees were first categorized into 'value-added', 'non-value-added', and 'waste' activities, based on the CIM (Common Information Model) standard classification. While the activities of 'value-added' (VA) activities refer to the work on the administrative product, for which the customers are willing to pay, the 'non-value-added' (NVA) activities refer to the ones that need to be done and the customers will not be paying for them. Hence, efforts and

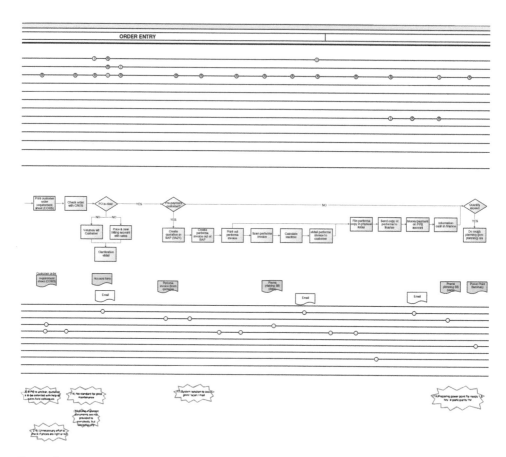

Figure 10.4 Evaluation of the As-Is process (Part 2).

time devoted to carrying out the concerned activities need to be reduced. 'Waste' (W) activities refer to neither administrative work nor the ones yielding financial gains, such as waiting time, searching, correction, etc., and therefore need to be reduced to the minimum. The survey conducted in Company X shows only 23% of FTE spent on VA activities while a significant portion (77%) is spent on NVA and W activities (51% and 26%, respectively), which do not contribute to any financial gains for the company (Figure 10.5).

This data on the three classified activities was on the core processes carried out in Company X involving the O2C process along with the methods of concept to launch, contact to order, general administration, general tasks, production/planning, and purchase to pay. Of these, the general tasks—the tasks which cannot be directly allocated clearly to specific processes (searching, waiting time, creating a report, etc.) and O2C process depict having substantial W activities, followed by purchase to pay process (Figure 10.6).

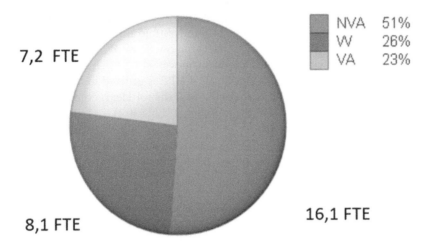

7,2 FTE

NVA 51%
W   26%
VA   23%

16,1 FTE

8,1 FTE

Figure 10.5 FTE spent on activities across core processes.

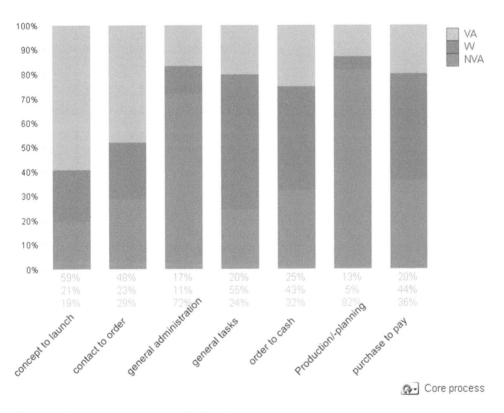

Figure 10.6 Segregation of classified activities across core processes.

In terms of FTE spent, W activities have a larger share in the O2C process over VA and NVA—1.8, 1.1, and 1.3 FTE, respectively. The process was further broken down into task categories and work items to understand the amount of FTE spent on the exact W activities. Out of 31.5 FTE, carrying out some W task categories utilizes about 8.7 FTE, such as clarification, handling, movement, searching, troubleshooting, wait, etc. 3.4 FTE goes to clarification, and 2.4 FTE is utilized in deposit and finding documents, which reflect a need for better standards for clear, complete, and precise information and its management in the company (Table 10.1).

Table 10.1 Identification of classified activities per task categories and work items

Task category	Number of FTEs	Percentage distribution (%)	Work item	Number of FTEs	Percentage distribution (%)
Clarification	3.4	10.8	Document	3.4	10.8
Fulfilment	3.3	10.4	Employee(s)	3.0	9.6
Recording	2.6	8.3	Report	2.7	8.7
Diverse	2.6	8.1	Diverse	2.5	8.0
Planning	2.2	7.0	Product	2.3	7.4
Handling	1.9	5.9	Auxiliary means	1.6	5.1
Control	1.9	5.9	Purchase order	1.5	4.7
Evaluation	1.3	4.1	Order	1.4	4.6
Movement	1.2	3.9	Standards	1.3	4.3
General administration	1.2	3.8	Ways	1.2	3.9
Searching	1.1	3.4	Master data	1.1	3.6
Internal service	1.0	3.1	Material	1.1	3.5
Care of business relationship	0.9	2.7	Hardware	0.9	2.8
Troubleshooting	0.8	2.7	Travel/visits	0.9	2.8
Training	0.7	2.4	Company	0.8	2.5
Development	0.5	1.7	Project	0.6	2.0
Maintenance	0.5	1.7	Software	0.5	1.5
Guide	0.5	1.6	Receivables	0.4	1.4
Inform	0.5	1.5	Responsibility	0.4	1.3
Order and cleanliness	0.5	1.5	Cost	0.4	1.1
Strategy	0.5	1.5	Data	0.3	1.1
Organization	0.5	1.5	Delivery note	0.3	1.0
Document maintenance	0.4	1.2	Price	0.3	1.0
Documentation	0.3	1.1	Purchase requisition	0.3	0.9
Wait	0.3	0.8	Organization	0.3	0.8
Fabrication	0.3	0.8	Supplier	0.2	0.8
Inspection	0.2	0.7	Debts	0.2	0.7
Standardize	0.2	0.5	Service provider	0.2	0.7
Research	0.2	0.5	Date	0.1	0.7
Installation	0.1	0.4	Quotation	0.1	0.4
Audit	0.1	0.3	Customer	0.1	0.3
Marketing	0.0	0.1	Asset	0.1	0.2
Acquisition	0.0	0.0	Event	0.0	0.0

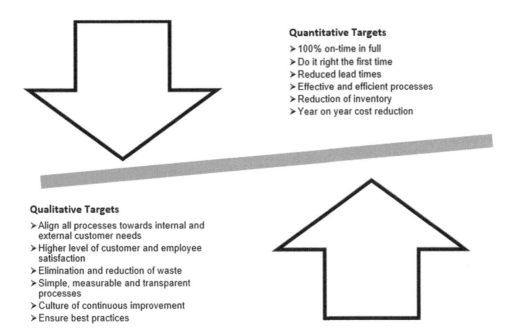

Quantitative Targets

➤ 100% on-time in full
➤ Do it right the first time
➤ Reduced lead times
➤ Effective and efficient processes
➤ Reduction of inventory
➤ Year on year cost reduction

Qualitative Targets

➤ Align all processes towards internal and external customer needs
➤ Higher level of customer and employee satisfaction
➤ Elimination and reduction of waste
➤ Simple, measurable and transparent processes
➤ Culture of continuous improvement
➤ Ensure best practices

Figure 10.7 Targets to enhance overall functioning.

10.3.3   Phase III: setting targets and defining future-state map

Initially, to enhance Company X's overall functioning through waste reduction, targets were developed considering the core elements of people, quality, costs, delivery performance, and customer satisfaction. The targets were qualitative and quantitative, based on employee empowerment and development principles, continuous flow, quality in product and processes, standards, synchronization, and customer orientation. The qualitative targets comprised aligning all processes towards internal and external customer needs; driving a higher level of customer and employee satisfaction; eliminating and reducing waste; maintaining simple, measurable, and transparent processes; ensuring a culture of continuous improvement; and ensuring best practices. The quantitative targets involved doing it right for the first time; being 100% on time in full; and reducing lead times, providing effective and efficient processes, reducing inventory, and reducing year-on-year costs (Figure 10.7).

Implementation of the action plan based on the targets set for the future state of the O2C witnessed the reduction of 30 process steps, additional reduction of the process by 17 documents, and complete utilization of the SAP – enterprise resource planning tool – benefits to avoid manual steps and get rid of unnecessary paperwork.

10.4   RESULTS AND INFERENCES

Based on the suggestion of the Kaizen bursts and waste activities identified above, 35 action items out of 33 measures were taken, of which nine quick wins were obtained,

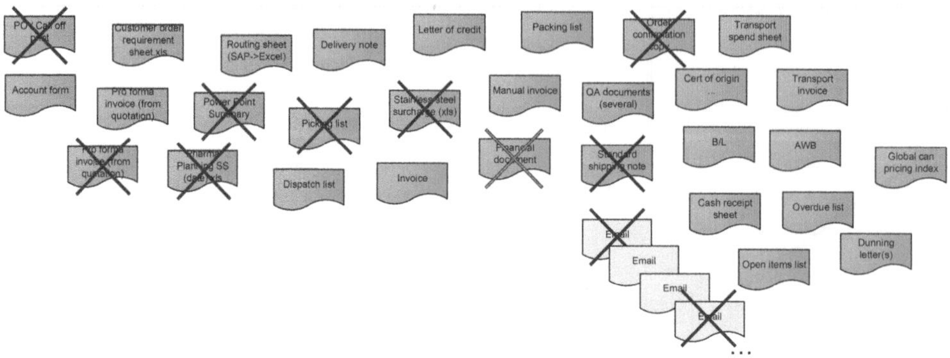

Figure 10.8 Cleaned up future state of the O2C process.

with the rest being achieved gradually over 12 months. Clear responsibilities were defined among the concerned employees to ensure the action plan's success, and the company was required to conduct regular follow-ups on the progress of each of these action plans. The nine quick wins obtained helped reduce waste activities of manual processing, paperwork, and overprocessing. For instance, enabling SAP to directly send documents (attachments) to recipients instead of scanning them and emailing them to forward further has addressed the waste activities of reduction in manual work, unnecessary paperwork, waiting time, and time for sorting at the printer. Avoiding manual credit limit adjustments for pre-payment customers led to a reduction in overprocessing (multiple documents carrying the same information), overproduction, and waiting time. Third, managing documents through the implementation of SAP and the DMS and reducing the existing number of records by making available only the useful ones through mergers and digitalization. Figure 10.8 shows the type of documents eliminated to clear the waste activities in document management.

The nine quick wins in the O2C process further highlighted the possibility of raising efficiency by 20–25% upon implementing lean techniques and consequently leading to more standardized processes, less clarification, system automatisms, and paperless work to name a few. When the assumed waste reduction (ranging from 15% to 50%) was applied to the FTE spent in each of these nine action items stated by the employees during the ABA, there was a considerable reduction in potential FTE. For instance, the employees invested nearly 3.20 FTE (considerably the highest among the waste activities) in clarification owing to the absence of standard information and processes. After lean admin implementation, in the future state, the assumed reduction in FTE was 25%, leading to an investment of potential FTE of 0.80. Following such an implementation, a substantial reduction (50%) in FTE invested is assumed to be in sending documents, lessening the previous FTE from 0.60 to 0.30. Cumulatively, 2.21 FTE is considered to be saved due to raising the process efficiency, which can further be invested in more value adding activities (Table 10.2).

While process steps decreased considerably by 30, documents to be managed have also come down to 47 from 64. Interestingly, Kaizen shows only four mini-improvement areas, down from 48 reflecting substantial effort being diverted to address the issues, which is further evident from the 33 measures being taken in the future-state

Table 10.2 Potential FTE reduction per activity after future state implementation

Category	FTE from ABA	Reduction in percentage	Potential FTE
Clarification	3.20	25	0.80
Order data entry	0.20	20	0.04
Rough planning	0.06	30	0.02
Credit limit	0.02	30	0.01
Invoice clearing	0.13	25	0.03
Searching	0.88	15	0.13
Sending	0.60	50	0.30
Paper handling	0.34	40	0.13
Meetings/other communication	2.99	25	0.75

Table 10.3 Final numbers after implementation of agreed measures

Count category	As-Is process	Future-state process
Process steps	93	63
Documents	64	47
Kaizens	48	4
Measures		33
Total saved efforts	2 FTEs	

process—an absent feature in the former. Lastly, all these steps resulted in a saved effort of nearly 2 FTEs ready to be invested in other value-added activities. Thus, observations from the three stages this study had embarked upon make the success of operational excellence in the O2C process evident. Implementing lean-based strategies with regular follow-ups and 'management for daily improvement' in the last 12 months post-workshop effectively helped save effort in FTEs and invest them in developing value-added activities (Table 10.3).

10.5 CONCLUSION

The findings resonate with the observations [4] that lean practices improve business performances, which is evident from reducing waste activities of process steps, documents, Kaizen bursts, and saved efforts in the form of FTEs. Implementation of lean practices helped Company X transform its waste activities into value-added ones in the O2C process. Through the usage of Kaizen bursts, the lean practices have helped the case company in waste identification and their elimination through systematic integration of all functions of the O2C process, thereby contributing to gaining efficiency [5] in the process. The contribution of Kaizen bursts in the identification of waste and its resolution through the implementation of IT tools fill the gap in research on the role of the concerned lean tools. Nonetheless, the finding supports the observations of Ghobakhloo & Hong [4] on the positive relationship between IT investments made by an organization concerning lean practices and improvement in business performance.

Overall, in terms of addressing the research gap in the academic field, this study brought forth the implementation of lean practices in the pharmaceutical component manufacturing industry, albeit in its initial stages. The approach enabled the industry (as is evident from the case study) to address problems in its O2C process like parallelism of KPIs, reduction of Kaizen bursts (improvement areas) from 48 to 4, standardization of processes, less clarification, system automatisms, and paperless work, to name a few. Employee cooperation in the implementation of lean practices in addressing the issues identified by Kaizens further helped reduce 2 FTES, which can be invested in value-added activities. The contribution of this study lies in bridging the existing gap in lean administration implementation in terms of analysing the present (As-Is) and developing the future (To-Be) state in the O2C process of the pharmaceutical component manufacturing company, using the lean techniques such as swim lane maps and realizing targets through Kaizen bursts. Additionally, through a phase-based methodology, this study contributes to understanding the pre-and-post implementation of lean management, establishing the technique's effectiveness in bringing down waste activities and diverting the effort towards value-added activities.

Therefore, the implication of the study lies in establishing the O2C operational efficiency and effectiveness through the recommended strategies presented in the future-state process. The study presents ways to manage documents, bring down the paperwork, the importance of standard information to save effort spent in clarifications, parallel working among departments, understanding of the process, and techniques to use in the smooth functioning of the O2C process. These implications are restricted to the O2C process of the pharmaceutical industry and other industries, especially the start-ups and Small & Medium Enterprises (SMEs), which being in the developed stage, might face problems maintaining a standardized work process.

The second implication of the study involves the establishment of the further scope of improvement in the lean administration of both Company X and its O2C process through the appointment of the process mentor. Hence, a future idea is provided to keep maintaining the benefits of lean in addressing waste activities and increasing efficiency. Therefore, through investigating the present case, the study has provided a road map to the top management of the case company and others in the industry to increase their operational efficiency, thereby making the imperativeness of lean management technique evident.

The third implication is that the case company underwent a major change in its processes—from an unorganized and unstandardized one, with considerable dependency on manual work—to a structured and standardized one through lean implementation. The findings also suggest the success of such an implementation. Often, the success or barriers associated with the implementation of lean techniques is dependent on the human-oriented softer factors like leadership and qualification of the employees. And a major change in the work environment often leads to employee resistance, as often such change requires employees to be qualified to handle the change processes. Here, lean implementation required the employees to be qualified to run the changed processes, especially the complete functioning of SAP, to reduce the effort invested in waste activities [10]. However, this study lacked the perspectives on challenges faced by the employees or the leaders (top management) in implementing lean and carrying out the processes to work on the Kaizen bursts. Future research capturing the perception of the employees and top management's perception of the challenges they faced while transforming the As-Is Swim Lane to future-state one while incurring the benefits or barriers can be carried out to bring forth strategies of minimizing the barriers of lean implementation.

REFERENCES

1. Jiang, Y. (2017). Prioritizing and selecting KPIs: translate performance results into managerial actions in strategy making process (Delft University of Technology).
2. Rohaan, D. (2017). Reengineering the order to cash (OTC) process (University of Twente). Retrieved from https://essay.utwente.nl/73224/1/Rohaan_BA_BMS.pdf.
3. Korotina, A., Mueller, O., & Debortoli, S. (2015). Real-time Business Process Intelligence. Comparison of different architectural approaches using the example of the order-to-cash process. *Wirtschaftsinformatik Proceedings*, 114, 1710–1724.
4. Ghobakhloo, M., & Hong, T. S. (2014). IT investments and business performance improvement: the mediating role of lean manufacturing implementation. *International Journal of Production Research*, 52(18), 5367–5384. Retrieved from https://www.tandfonline.com/doi/abs/10.1080/00207543.2014.906761.
5. Nenni, M. E., Giustiniano, L., & Pirolo, L. (2014). Improvement of manufacturing operations through a lean management approach: a case study in the pharmaceutical industry. *International Journal of Engineering Business Management Special Issue: Innovations in Pharmaceutical Industry*, 6(24), 6. DOI: 10.5772/59027.
6. Villarraga, J., Carley, K. M., Wassick, J., & Sahinidis, N. (2017). Agent-based modeling and simulation for an order-to-cash process using netlogo.
7. Moosivand, A., Ghatari, A. R., & Rasekh, H. R. (2019). Supply chain challenges in pharmaceutical manufacturing companies: using qualitative system dynamics methodology. *Iranian Journal of Pharmaceutical Research*, 18(2), 1103–1116. Retrieved from https://www.ncbi.nlm.nih.gov/pmc/articles/PMC6706717/.
8. Bevilacqua, M., Ciarapica, F. E., De Sanctis, I., Mazzuto, G., & Paciarotti, C. (2015). A Changeover Time Reduction through an integration of lean practices: a case study from pharmaceutical sector. *Assembly Automation*, 35(1), 22–34.
9. Rybski, C., & Jochema, R. (2016). Benefits of a learning factory in the context of lean management for the pharmaceutical industry. *Procedia CIRP*, 54, 31–34. Retrieved from https://www.sciencedirect.com/science/article/pii/S2212827116308824
10. Sieckmann, F., Nguyen Ngoc, H., Helm, R., & Kohl, H. (2018). Implementation of lean production systems in small and medium-sized pharmaceutical enterprises. *Procedia Manufacturing*, 21, 814–821. Retrieved from https://www.sciencedirect.com/science/article/pii/S2351978918302282.
11. Sampson, S. E. (2012). Visualizing service operations. *Journal of Service Research*, 15(2), 182–198.
12. Gallenkämper, J. (2016). Kaizen teaching and the learning habits of engineering students in a freshman mathematics course. *Central European Journal of Operations Research*, 24(4), 1009–1030. Retrieved from https://link.springer.com/article/10.1007/s10100-015-0416-5.
13. Dzenopoljac, V., Alasadi, R., Zaim, H., & Bontis, N. (2018). Impact of knowledge management processes on business performance: evidence from Kuwait. *Knowledge and Process Management: The Journal of Corporate Transformation*, 25(2), 77–87. Retrieved from https://onlinelibrary.wiley.com/doi/abs/10.1002/kpm.1562.
14. Hallam, C. R. A., Valerdi, R., & Contreras, C. (2018). Strategic lean actions for sustainable competitive advantage. *International Journal of Quality & Reliability Management*, 35(2), 481–509. Retrieved from https://www.emerald.com/insight/content/doi/10.1108/IJQRM-10-2016-0177/full/html.

Chapter 11

Modelling and analysis of Lean Six Sigma framework along with its environmental impact on the business process

A review

Deepak Sharma, Dharmendra Singh, Rahul Sharma, and Y. B. Mathur

CONTENTS

11.1 INTRODUCTION

The major aim of the study is to examine the modelling and analysis of the LSS framework along with its environmental impact on business processes. Each company's presentation depends on the way that it must consolidate fundamental advances that make it maintainable, use assets proficiently, and embrace pertinent administration styles that can enable it to develop. Such measures depend on picking different administration styles for the upgrade of firm execution, and for that reason, this examination centre around assessing three distinctive administration styles [1]. The three structures are recorded as Lean, Six Sigma, and natural supportability. Lean spotlights on squander decrease and disposing of exercises that don't increase the value of the business measures [2]. The common-sense pertinence of Lean has been helpful in the assembling and administration areas, including human services, banking, and instruction. Continuous development of Lean methods happens in Kanban and Kaizen with various associations of other platforms [3].

DOI: 10.1201/9781003293576-11

Six Sigma is a quality administration device broadly embraced by organizations to lessen varieties and deformities in their items and cycles. As of late, Six Sigma has been affirmed by associations, including money-related undertakings, clinics, instructive establishments, and electronic business [4]. Six Sigma is a very effective method to control the various steps involved in industries and help to accomplish the goal of the industries. For example in 1999, in the US, General Electric Company saved 2 billion US dollars by applying the Six Sigma methods in its plants [5].The same advantage is achieved by Motorola Company. The Lean version of Six Sigma is also more effective for industrial optimum development [6]. The current study shows that numerous frameworks or systems on Lean, Six Sigma as well as on LSS have been proposed by various scholars within the different settings. The current study proposed to fundamentally review the present LSS structures along with the frameworks more than a few boundaries like the curiosity, structure check, methods of structure confirmation, distinguishing between the imperative components/apparatuses/builds, relative investigation of all chosen structures and recommending potential rules for future assessment of LSS systems. As LSS is a crossover approach of Lean and Six Sigma systems, the creators have thought about structures of individual strategies in the current investigation.

Ecological supportability is tied in with using normal assets without trading off the earth for people in the future [7]. Also, biological mindfulness is a need considered for the entirety of the corporate divisions, enterprises and associations, and so on. Organizations, alongside their upper hands over different firms, should likewise guarantee to adjust their picture in the public eye by keeping the principles of ecological maintainability. To contend in the worldwide market, organizations must concentrate on natural supportability that can give them an upper hand over the long haul [8]. Zhu et al. studied the application of Six Sigma to manage the ecological system where industries are located [9]. Be that as it may, some examination works repudiate and guarantee a negative connection between ecological maintainability and firm execution. Their discoveries demonstrate that putting resources into maintainable practices increments their expenses as well as the final result costs which influences the money-related execution of the organization because of the decrease in benefits and piece of the overall industry [10–13]. In light of the conversation referenced before, the impact of ecological manageability on Small- and Medium-Sized Enterprises (SMEs) exhibition is yet to be surveyed with regards to a creating nation. The concept of Lean is the first in the world, introduced by Taichi Ohno of Toyota, to control the production delays and unwanted defects. Lean has various definitions appropriate in various situations [14]. The Lean idea plans to make and convey the items and administrations in manners that utilize the least expense and time, along these lines of diminishing waste. Essentially, it presently can't seem to reach out toward the creating nations where the obstruction of progress is extremely high and conventional work rehearses command [15]. Besides, Lean is centred on supportability to increase an upper hand for the firm. This is done to address the clients' issues and requests by planning work processes and asset distribution measures. It very well may be made conceivable by taking a gander at the assembling measures in the restricted assets circumstances [16]. To conduct literature review for this research study, some research steps are followed by researchers which are present in Figure 11.1 and Table 11.1.

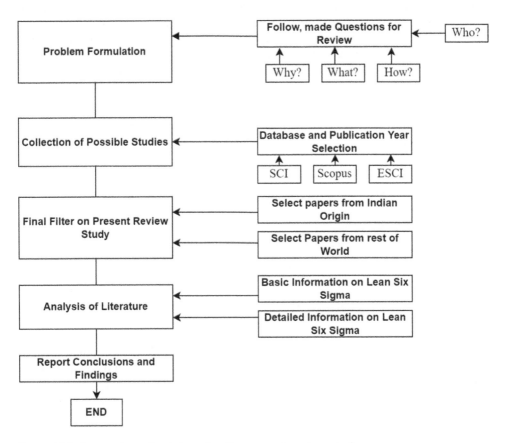

Figure 11.1 Research methodology for the present review study.

Table 11.1 Research criteria inclusion and exclusion for the present review paper

S. no.	Content	Inclusion	Exclusion
1	Database selection	SCI, Web of Science, Scopus, ESCI	Paid Journals, fake or clone Journals, Low-ranking Journals
2	Publication year	2003–2020	Before 1990
3	Keywords	Six Sigma, Lean, manufacturing, India, Service Sector, Manufacturing	Six Sigma in the medical industry
4	Thesis	Only Indian Thesis	Rest of the world

11.2 LITERATURE REVIEW

A literature review can be defined as a review article or paper which contains the recent and earlier findings, including methodological and theoretical efforts in a particular field or a particular topic. A literature review doesn't include current and new research or experiment; it is a secondary source having all the discussions and pieces of information from previous studies. This section provides an overview of recent and previous studies on the same or almost relevant topics including some subsections of it. This chapter also assesses and discusses integrative knowledge and strategies in order to develop a better understanding of the LSS framework along with the environmental impact on business processes.

11.2.1   Six Sigma and Lean manufacturing in the context of Sustainability

As of late, with the ascent of tasks, social, natural, and quality improvement approaches, for example, Lean, Green activities (hereinafter Green), Six Sigma, among others, and expanding worries for the earth and social duty, the market elements have transformed. Customarily, creation proficiency and benefit, and later adaptability, quality, and consumer loyalty developed as new serious models [17]. Notwithstanding, with the developing weight from different partners to improve social and ecological execution, associations have now been compelled to transform their styles in order to deal with the overseeing cycles and tasks [18–19]. Green et al. studied the importance of the three keywords– monetary, social, and ecological – for the development of the industry and its local ecological conditions [20]. In this situation, the test for associations is to address every one of their partners' issues through achieving positive financial execution while finding the correct parity among the triple main concerns of maintainability [21]. Various researchers combine the Lean and Green Lean Six Sigma (GLSS) methods in their research work and find the importance of these methods [22–28].

As discussed in Figure 11.2, it is an integrated view of Lean, Six Sigma, and Sustainability, whereas Six Sigma controls the process variation, Lean is utilized to control the wastage, and Sustainability is implemented to control the pollution. Hence,

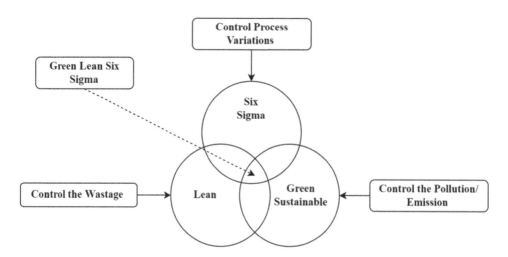

Figure 11.2 Lean, Six Sigma, and Sustainability.

Green Lean Six Sigma is the common outcome found among integration of Lean, Six Sigma, and Sustainability.

According to Cherrafi et al., 85.9% is showing the incorporation of Lean and Sustainability, whereas 8.5% is showing the compilation of Six Sigma and Sustainability. The common percentage of 5.6% is showing the GLSS, which is basically the mixture or emergence of these three having all three features. It helps in waste reduction, value addition, and resource conservation [22].

In the current worldwide scenario, the primary challenges and issues that our social orders face are environmental change, populace development, neediness and imbalance, contamination, as well as the expanding cost of vitality and assets [29]. Endeavours to improve natural and social manageability of modern cycles have customarily been seen as hindrances to the financial maintainability of an association [30,31], yet as of late numerous organizations have found that those endeavours bring about decreased working expenses and improved worker fulfilment [32,33]. Lately, the utilization of the executives' frameworks to fathom the current worldwide test of manageability has been investigated [34]. Lean and Six Sigma are capable to control the defects, unwanted time wastage, and management of the company in a better way [30–40].

11.2.2   Integrated impact of Green manufacturing and Lean Six Sigma (LSS) on the manufacturing industries' environmental performance

The idea of Lean got well known through the book "The Machine that Changed the World". Lean creation has been characterized from various perspectives [35]. One explanation behind the absence of a rational definition may be that the idea is as yet advancing. In any case, the main objective of a Lean framework is to create items and administrations of "higher calibre" at a highly decreased expense and at all times by killing squander [36]. The USA's Environmental Protection Agency (EPA) characterized the points of Lean as: "Build up the greatest items, at the most reduced expense, with the briefest lead time by deliberately and consistently disposing of waste, while regarding individuals and nature" [37]. According to Chiarini and Vagnoni, various methods are available in LSS, which are 5S (Sort, Set in Order, Shine, Standardize, Sustain), SMED (Single Minute Exchange of Die), TPM (Total Productive Maintenance), and Kanban. All these methods help to apply Lean in any type of industry [38]. Various researchers find that Lean and Six Sigma can be applied to the best knowledge or practice techniques. The DMAIC (Define–Measure–Analyze–Improve–Control) approach is also available for Six Sigma application as shown by various researchers [38,39]. According to Chiarini, from a useful perspective, experts named Black and Green Belts do problem-solving ventures utilizing quality and factual instruments from the TQM (Total Quality Management) world [40]. Lucato et al. try to manage the control among the environment and companies by using Six Sigma techniques. In their study, various important considerations are provided by these researchers [41]. Some other researchers are also trying to control the pollution due to industrial development by applying the Lean and Six Sigma methods [40–42]. Undoubtedly, the Six Sigma execution improved the progressive capacity for data to put together errands with respect to the board [42]. In their study conducted by Habidin et al., they provided the importance of Six Sigma to manage the ecological development with industrial development [43].

11.2.3 LSS framework advancements with time

Product quality and its production time are very important parameters for most of the industries of majority sectors. These production parameters are more feasible in the present era because today's small scale industry has also international customers. Due to the importance of the quality of the product, the industry is required to manage the quality of the product by using industrial tools which help to maintain the quality of the product. Various researchers like Vijaya Sunder [44] investigated the various tools to improve the quality of the product. In their research, they discuss the importance of the tools to maintain the quality of the product and its cost of application and yield on customers. All major issues are discussed in their research.

In recent years, the tools used for manufacturing sectors to improve their production quality are now used in various service sectors to improve the working culture of the organizations. Of all industrial tools, LSS is the most useful and applicable for almost all industries and organizations. LSS follows the continuous development of the framework for most industries and proves its importance for maintaining its quality [45]. Various researchers like Simons show the benefits of LSS over other industrial tools like TQM and Lean. LSS proves its importance over other tools used by industries [46]. Tolerating the significant possibility, a framework for LSS association has been created in various sectors for setting out the quality importance for the product [47]. A productive six steps based sensible method is suggested in various papers [17]. As seen in Figure 11.3, the most advanced development of LSS is Strategic LSS Framework. This framework is more capable to make better relations between Lean and Six Sigma methods. It means Lean can use easily the Critical to Quality (CTQ) tools in its application by using this framework [48].

Lean and Six Sigma industrial tools have been noticeable among scholastics and specialists for a long time. The Lean cycle remembers five vital standards for which the authority is directed towards execution. These rules (a) specify value according to the client viewpoint; (b) align within worth stream with what the client respects; (c) make the stream; (d) pull on request; and (e) make impeccability [49]. The most applicable method available in Six Sigma is the DMAIC method which has five steps to complete the Six Sigma cycle and more useful tool for most industrial sectors. D represents the design of the problem, M represents the measure, A represents the analysis of the data, I represents the improvement in data, and C represents the control of the system [50]. Endeavours to work with Lean and Six Sigma methods both have a bound together methodology which helps to improve the Lean and Six Sigma for industrial development by and large developed using the most useful tools of Six Sigma which is called DMAIC.

All things considered, when seeing the degree of scholarly and expert making around the utilization out of LSS undertakings, generally scarcely any splendid lights on the Lean execution cycle to any huge degree. In the present chapter, an improved framework of the LSS cycle is used, one which puts the Lean speculation at the focal point of the cycle endeavouring to absolutely mistreat the positive conditions that both Lean and Six Sigma can accommodate connections.

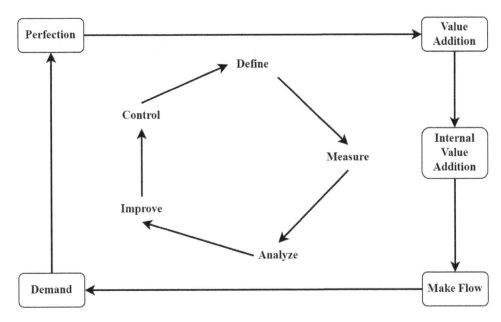

Figure 11.3 Strategic LSS Cycle.

Two or three Six Sigma instruments and methods are used for the improvement of the material idea of the board [51]. Hill et al., in their paper, used an improved framework of LSS to control the production issues and succeeded to apply this method in their research work. The proposed LSS structure was made for gathering relationships to give synergic benefits and further develop purchaser dependability. The proposed model was realized in a couple of collecting affiliations and its ampleness was surveyed [52]. Various other researchers like Cheng and Chang, and Vinodh et al. also studied the LSS framework to improve the production and service quality of the industry and organizations. Besides, it was found that using DMAIC structure with LSS instruments and techniques, for instance, adventure contract, the load up exercises, activity characterization, waste course of action, Pareto assessment, and reasonability plan for decreasing disfigurements [53,54].

Kumar et al. studied the application of the LSS framework for Indian small scale industries and found better results by using this LSS method, especially the application of DMAIC is the most relevant tool for Indian sectors. The objective was to diminish leaves for gathering lines and to further develop customer dedication with convincing improvement in key estimations [55] (Figure 11.4).

The above image defines the working of LSS in order to move towards sustainable development. First, the problem is defined, and then to measure the problem and its consequences, some data collection has to be done. Moving forward, analysis takes place and includes critical factors. The second last step in this process is to identify the different improvement strategies and techniques. The last step is to make moves in order to control the improvements achieved.

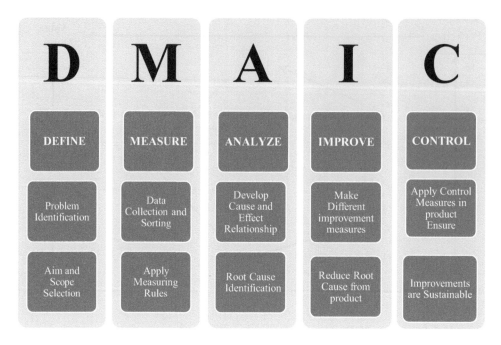

Figure 11.4 Working steps for LSS to move forward with sustainable development (DMAIC approach).

11.3 FINDINGS AND DISCUSSIONS

The findings for the current study were gained from the secondary data collected. To fulfil the study objectives, a qualitative secondary research approach has been undertaken. There have been many studies on strategic planning for start-ups, but this study specifically targets the context of the LSS framework with respect to business processes in an environmental context. For endurance of any association in the Lean and Six Sigma can help to find the consumer problems with industry and its products by using the most effective method, i.e., the DMAIC approach in Six Sigma.

Numerous enterprises have represented colossal additions, while a few associations have not gotten the ideal outcomes. One reason for not achieving the genuine advantages of LSS by the businesses is that the administration, chiefs, and workers don't perceive what comprises LSS.

It is very much difficult for any organization to work within all environmental measures considering their own manufacturing or any other firm by providing utilization opportunities and profits for the same. Hence, it is very important for start-ups to adopt strategic planning and various techniques in order to survive in a market where competition is much higher. The study also examined most of the LSS strategies in order to guide the organization for better and more effective business processes considering good and healthy environmental situations as well.

11.4 CONCLUSION AND RECOMMENDATIONS

The point of this examination is to inspect the province of LSS research. Consistently, numerous creators propose structures for Lean, Six Sigma, and LSS. As presented in this review, the application of Green Lean Six Sigma and Lean assembly can improve the working of the industry and also help to control the ecological conditions. Scopus information base was utilized for the underlying hunt of diaries. The LSS system introduced here is so exceptionally successful and useful for associations and firms to use so as to work out positively for the earth. In this chapter, a few scientific categorizations were built up for looking into systems and they were grouped in. Significant structures are novel, which shows great development in new fields. The components/builds utilized in the advancement of these structures are ambiguous. Different structures are profoundly unique or shallow. A portion of the structures doesn't show how each build/component is identified with execution in the association. Setting up a business and developing it at a colossal level, for the high benefit, affirmation, and administration giving capacities, is consistently hard for an organization. All things considered, the association of the firm is needed to go deliberately, the primary thing is that surfaces are arranging a proficient and significant methodology so as to develop their organization and have high efficiency. This investigation shows different inadequacies of the created structure. The momentum study will support different researchers and analysts to direct their elective investigations in a deliberate and sensible way. The results structure of the investigation will likewise assist them with building up their one and one-of-a-kind ideas for a similar zone of premium. Besides, the writing will be useful for the organizations and associations to well develop their business measures with regard to ecological effects.

REFERENCES

1. Garza- Reyes, J.A., Al-Balushi, M., Antony, J., andKumar, V. (2016). A Lean Six Sigma Framework for the reduction of ship loading commercial time in the iron ore pelletising industry. *Production Planning & Control*, 27(13), 1092–1111.
2. Schulze, A., and Stormer, T. (2012). Lean Product Development –enabling management factors for waste elimination. *International Journal of Technology Management*, 57(1–3), 71–91.
3. Jordan, T. (2017). Kaizen Kanban: a visual facilitation approach to create prioritized project pipelines. *Quality Progress*, 50(6), 69.
4. Alexander, P., Antony, J., and Rodgers, B. (2019). Lean Six Sigma for small-and-medium-sized manufacturing enterprises: a systematic review. *International Journal of Quality and Reliability Management*, 36(3), 378–397.
5. Ali, Y., Younus, A., Khan, A.U., and Pervez, H. (2020). Impact of Lean, Six Sigma and environmental sustainability on the performance of SMEs. *International Journal of Productivity and Performance Management*, 70(8), 2294–2318.
6. Singh, M.,and Rathi, R. (2019). A structured review of Lean Six Sigma in various industrial sectors. *International Journal of Lean Six Sigma*, 10(2), 622–664.
7. Bauman, S. (2015). *What is Sustainability?* UNT Health Science Centre, Fort Worth, TX.
8. Lopez, M.V., Garcia, A., and Rodriguez, L. (2007). Sustainable development and corporate performance: a study based on the Dow Jones sustainability index. *Journal of Business Ethics*, 75(3), 285–300.

9. Zhu, Q., Johnson, S., and Sarkis, J. (2018). Lean Six Sigma and environmental sustainability: a hospital perspective. *Supply Chain Forum: An International Journal*, 19(1), 25–41.

10. Brammer, S., and Millington, A. (2008). Does it pay to be different? An analysis of the relationship between corporate social and financial performance. *Strategic Management Journal*, 29(12), 1325–1343.

11. Friedman, M. (2007). *The Social Responsibility of Business is to Increase its Profits. Corporate Ethics and Corporate Governance*, Springer, Berlin, Heidelberg, pp. 173–178.

12. Tang, Z., Hull, C.E., and Rothenberg, S. (2012). How corporate social responsibility engagement strategy moderates the CSR-financial performance relationship. *Journal of Management Studies*, 49(7), 1274–1303.

13. Williams, H.E., Medhurst, J., and Drew, K. (1993). Corporate strategies for a sustainable future. *Environmental Strategies for Industry: International Perspectives on Research Needs and Policy Implications*, 117, 146.

14. Hadded, M., and Otayek, R. (2018). *Addressing the challenge of Lean Manufacturing Sustainment with System Dynamics Modeling: A Case Study on Apprel Manufacturing in a Developing Country.* Lebanese American University, Paris.

15. Douglas, J., Muturi, D., Douglas, A., and Ochieng, J. (2017). The role of organisational climate in readiness for change to Lean Six Sigma. *The TQM Journal*, 29(5), 666–676.

16. Ulewicz, R., and Kuceba, R. (2016). Identification of problems of implementation of Lean concept in the SME sector. *Ekonomiai Zarzadzanie*, 8(1), 19–25.

17. Garza-Reyes, J.A. (2015). Green Lean and the need for Six Sigma. *International Journal of Lean Six Sigma*, 6(3), 226–248.

18. Wong, W.P., and Wong, K.Y. (2014). Synergizing an ecosphere of lean for sustainable operations. *Journal of Cleaner Production*, 85(15), 51–66.

19. McCarty, T., Jordan, M., and Probst, D. (2011). *Six Sigma for Sustainability – How Organizations Design and Deploy Winning Environmental Programs.* McGraw–Hill, New York.

20. GreenJr, K. W., Zelbst, P. J., Meacham, J., and Bhadauria, V. S. (2012). Green supply chain management practices: impact on performance. *Supply Chain Management: An International Journal*, 17(3), 290–305.

21. Alves, J. R. X., and Alves, J. M. (2015). Production management model integrating the principles of lean manufacturing and sustainability supported by the cultural transformation of a company. *International Journal of Production Research*, 51(11), 1–14.

22. Cherrafi, A., Elfezazi, S., Chiarini, A., Mokhlis, A., and Benhida, K. (2016). The integration of lean manufacturing, Six Sigma and sustainability: aliterature review and future research directions for developing a specific model. *Journal of Cleaner Production*, 139, 828–846.

23. Carvalho, H., and Cruz-Machado, V. (2009). Lean, agile, resilient and green supply chain: areview. In *Proceedings of the Third International Conference on Management Science and Engineering Management*, Bangkok, 2 November, 3–14.

24. Franchetti, M., Bedal, K., Ulloa, J., and Grodek, S. (2009). Lean and green: industrial engineering methods are natural stepping stones to green engineering. *Industrial Engineer*, 41(9), 24–29.

25. Bergmiller, G. G., and McCright, P. R. (2009). Are lean and green programs synergistic? *Industrial Engineering Research Conference*, Miami, FL, May 30–June 3.

26. Dües, C. M., Tan, K. H., and Lim, M. (2013). Green as the new lean: how to use lean practices as a catalyst to greening your supply chain. *Journal of Cleaner Production*, 40, 93–100.

27. Martinez-Jurado, P. J., and Moyano-Fuentes, J. (2014). Lean management, supply chain management and sustainability: aliterature review. *Journal of Cleaner Production*, 85, 134–150.

28. Hajmohammad, S., Vachon, S., Klassen, R. D.,and Gavronski, I. (2013).Lean management and supply management: their role in green practices and performance. *Journal of Cleaner Production*, 39, 312–320.

29. Sreedharan, V. R., Pattusamy, M., Mohan, S., and Persis, D. J. (2020). A systematic literature review of Lean Six Sigma in financial services: key finding and analysis. *International Journal of Business Excellence*, 21(3), 331–358.

30. Khor, M. (2011). Challenges of the green economy concept and policies in the context of sustainable development, poverty and equity. *The Transition to a Green Economy: Benefits, Challenges and Risks from a Sustainable Development Perspective*, 69–97.

31. Wilson, A. (2010). *Sustainable Manufacturing: Comparing Lean, Six Sigma, and Total Quality Manufacturing*. Strategic Sustainability Consulting, Washington, DC.

32. Corbett, C. J., andKlassen, R. D. (2006). Extending the horizons: environmental excellence as key to improving operations. *Manufacturing & Service Operations Management*, 8(1), 5–22.

33. Simboli, A., Taddeo, R., and Morgante, A. (2014). Value and wastes in manufacturing. An overview and a new perspective based on eco-efficiency. *Administrative Sciences*, 4(3), 173–191.

34. Chiarini, A. (2015). Environmental policies for evaluating suppliers' performance based on GRI indicators. Sustainable manufacturing-greening processes using specific Lean Production tools: an empirical observation from European motorcycle component manufacturers. *Journal of Cleaner Production*, 85, 226–233.

35. Dahlgaard, J. J., and Mi Dahlgaard-Park, S. (2006). Lean production, six sigma quality, TQM and company culture. *The TQM Magazine*, 18(3), 263–281.

36. Dennis, P. (2017). *Lean Production simplified: A Plain-Language Guide to the World's Most Powerful Production System*. CRC Press, Boca Raton, FL.

37. EPA (2003). *Lean Manufacturing and the Environment: Research on Advanced Manufacturing Systems and the Environment and Recommendations for Leveraging Better Environmental Performance*. United States Environmental Protection Agency.

38. Chiarini, A., and Vagnoni, E. (2014). World-class manufacturing by Fiat. Comparison with Toyota production system from a strategic management, management accounting, operations management and performance measurement dimension. *International Journal of Production Research*, 53(2), 590–606.

39. Linderman, K., Schroeder, R., Zaheer, S., and Choo, A. (2003). Six Sigma: agoal theoretic perspective. *Journal of Operations Management*, 21(2), 193–203.

40. Chiarini, A. (2013). Building a Six Sigma Model for the Italian public healthcare sector using grounded theory. *International Journal of Services and Operations Management*, 14(4), 491–508.

41. Lucato, C.W., Junior, V.M., andSantos, S.D.C.J. (2015). Eco-Six Sigma: integration of environmental variables into the Six Sigma technique. *Production Planning & Control*, 26(8), 605–616.

42. Calia, C.R., and Castro de, M.G.M. (2009). The impact of Six Sigma in the performance of a Pollution Prevention Program. *Journal of Cleaner Production*, 17(15), 1303–1310.

43. Habidin, F. N., and Yosof, M.S. (2012). Relationship between lean six sigma, environmental management systems and organizational performance in the Malaysian Automotive Industry. *International Journal of Automotive Technology*, 13(7), 1119–1125.

44. Vijaya Sunder, M. (2016). Constructs of quality in higher education services. *International Journal of Productivity and Performance Management*, 65(8), 1091–1111.

45. Pepper, M.P.J., and Spedding, T.A. (2010). The evolution of Lean Six Sigma. *International Journal of Quality & Reliability Management*, 27(2), 138–155.

46. Simons, N. (2013). The business case for Lean Six Sigma in higher education. *ASQ Higher Education Brief*, 6(3), 1–6.

47. Antony, J., and Sunder, M.V. (2018). A conceptual Lean Six Sigma framework for quality excellence in higher education institutions. *International Journal of Quality & Reliability Management*, 35(4), 857–874.
48. Thomas, J.A., Francis, M., Fisher, R., and Byard, P. (2016). Implementing Lean Six Sigma to overcome the production challenges in an aerospace company. *Production Planning & Control*, 27(7–8), 591–603.
49. Womack, J. P., Jones, D. T., & Roos, D. (2007). The Machine That Changed the World: *The Story of Lean Production-Toyota's Secret Weapon in the Global Car Wars That Is Now Revolutionizing World Industry.* Simon and Schuster.
50. Harry, M., and R. Schroeder. (2006). *Six SIGMA: The Breakthrough Management Strategy Revolutionizing the World's Top Corporations.* Dell Publishing, New York.
51. Chen, M., and Lyu, J. (2009). A Lean Six Sigma approach to touch panel quality improvement, *Production Planning & Control*, 20(5), 445–454.
52. Hill, J., Thomas, A. J., Mason-Jones, R. K., and El-Kateb, S. (2018). The implementation of a Lean Six Sigma framework to enhance operational performance in an MRO facility. *Production & Manufacturing Research*, 6(1), 26–48.
53. Cheng, Y.C., and Chang, Y.P. (2012). Implementation of the lean six sigma framework in non-profit organisations: acase study. *Total Quality Management & Business Excellence*, 23(3–4), 431–447.
54. Vinodh, S., Gautham, S.G., and Ramiya, R.A. (2011). Implementing Lean Sigma framework in an Indian automotive valves manufacturing organisation: a case study. *Production Planning & Control*, 22(7), 708–722.
55. Kumar, M., Antony, J., Singh, R.K., Tiwari, M.K., and Perry, D. (2006). Implementing the lean six sigma framework in an Indian SME:acase study. *Production Planning & Control*, 17(4), 407–423.

Chapter 12

Optimum order allocation in a multi-supplier environment using linear programming model

Case study on heavy industry in India

K. Muthu Narayanan, R. S. Aakhash, K. Dharun Prashanth, and P. Parthiban

CONTENTS

12.1 INTRODUCTION

A supply chain is an integrated network of various processes and entities to deliver the final product in the desired quantity and quality to the end customer at the right place at the right time. It includes not only manufacturers and suppliers but also transporters, warehouses, retailers, and even customers. It involves the integration of these business entities to satisfy the customers' needs and demands efficiently. The supply chain forms the channel for the flow of material and information. The business processes included in a supply chain may be classified as supply/inbound logistics, production and distribution, and outbound logistics. The various activities involved in the supply chain range from sourcing to customer servicing. It also includes the crucial components of coordination and collaboration with channel partners, which can be suppliers, intermediaries, third-party service providers, and customers. In essence, supply chain management integrates supply and demand management within and across companies.

Sourcing is basically defined as *the set of business processes required to purchase goods and services.* Sourcing decisions are basically operational level decisions taken in a supply chain which are crucial because sourcing affects the level of efficiency and responsiveness of the supply chain. Supplier Selection is one of the key stages of supply chain management (SCM) because the cost of raw materials and components' parts constitutes the main cost of a product.

The objective of supplier selection in a supply chain is to identify suppliers with the highest potential for meeting a firm's needs consistently and at a mutually acceptable

DOI: 10.1201/9781003293576-12

cost (Zeydan et al., 2011). Several factors may affect a supplier's performance and therefore supplier selection is a multi-criteria problem which includes both tangible and intangible criteria (Weber et al., 1991). Some of these criteria may conflict with each other. Many techniques have been proposed and implemented for vendor evaluation and selection; most of them try to rank the vendors from the best to the worst and choose the appropriate vendor(s).

The next vital stage in the sourcing process after choosing the suitable suppliers is Order Allocation. This stage is required to determine the optimal materials quantities to be obtained from each supplier, especially in a multiple-supplier environment. In a multiple sourcing environment, no supplier will be able to satisfy all the buyer's requirements; hence, there is a requirement of more than one supplier to be identified for order allocation. In this situation, the buyer purchases the same items from more than one supplier, and the total demand is split among them (Pan, 1989). The order quantity is split among suppliers for a variety of reasons such as cost, quality, and capacity. This decision can be seen in cases where the supplier selection and the order allocation problems are combined (Turner, 1988). In this decision, the following queries are being addressed: What order quantity should be allocated to each supplier? Which order should be assigned to each supplier? And which period, in the planning horizon, should be used? Various techniques have been developed to solve the optimal order allocation problem, including linear programming, non-linear programming, mixed-integer programming, and artificial intelligence technique (Rosenthal et al., 1995). These methods mainly focused on single-objective optimization, that is, minimizing cost. However, in a real-world supply chain environment, the decision-maker must consider uncertain factors along the supply chain. To reduce the risk, multi-objective optimization models are being identified to strike a trade-off between two or more conflicting objectives involved in the order allocation process.

12.2 LITERATURE REVIEW

In the case of multiple sourcing, where no single supplier will be able to meet all the buyer's requirements, more than one supplier needs to be selected for ordering. This stage is termed the Order Allocation Stage. Several techniques have been developed for solving the optimum order allocation problem. Some of the techniques include linear programming, non-linear programming, mixed-integer programming, and artificial intelligence technique.

In general, these order allocation problems will focus on an objective function which has to be either maximized or minimized. This objective function will be subjected to several constraints. In the past, most of the order allocation problems will deal with only single objective, i.e., to minimize the cost or to maximize the total value of purchasing. Ghodsypour and O'Brien (1998) proposed a Decision Support System for supplier selection and order allocation by integrating Analytic Hierarchy Process (AHP) and linear programming. In this linear programming model, the objective was to maximize the total value of purchasing subjected to Capacity, Quality, and Demand constraints. They had solved the model with real-time data from a JIT manufacturer and identified the optimum quantity to be ordered from the four suppliers.

In another research conducted by Jayaraman et al. (1999), minimizing the fixed and variable costs for the organization was the only objective subjected to main constraints like Quality, Storage, and Production Capacity. Also, constraints like Lead Time, Supplier Count, Spare Capacity, and Non-negativity constraints were used in the model.

However, in the present world supply chain environment, more than one objective are being considered for the order allocation problems. Many research works have proposed multi-objective optimization models. Liu et al. (2013) studied a Logistics Service Supply Chain (LSSC) consisting of one Logistics Service Integrator (LSI) and many Functional Logistics Service Providers (FLSPs). They had created a model having two goals, minimizing the cost of the LSI and maximizing the subjective utilities of FLSPs. MATLAB® 7.8 was used to solve a numerical example of the above model.

In a research conducted (Razmi et al., 2009), an integrated fuzzy approach for supplier selection and order allocation was proposed. In order allocation, four objective functions were used, namely minimizing the cost, delivery lateness, and defective product and maximizing the utility function subjected to constraints like Capacity, Demand, and Product Balance. The proposed model was implemented in a car production industry in Iran.

Further, Faez et al. (2009) proposed an integrated fuzzy case-based reasoning and linear programming model where three objective functions were discussed. The objective of their model was to minimize the total cost, minimize the total rejections, and maximize the on-time deliveries. The constraints proposed here were demand and capacity constraints.

Nazari-Shirkouhi et al. (2013) had proposed a supplier selection and order allocation problem under multi-price level and multi-product using an interactive two-phase fuzzy multi-objective linear programming model. The model attempted to simultaneously minimize the four components taken as individual objectives, namely total purchasing costs, ordering costs, number of defective units, and number of late deliveries. Their model and numerical results proved that this approach is highly effective in an uncertain environment and proves to be a reliable decision-making tool.

12.3 METHODOLOGY

Order allocation problems help us to identify the optimal load to be allocated to a supplier in a given specific environment and conditions. When there is no single supplier who will be able to meet the entire requirement of the manufacturer reasons being capacity, lead time, demand of the product, etc., there is a need for identification and selection of more than one supplier. In such cases, optimum load allocation helps us to divide the demand amongst the available suppliers accordingly to achieve the desired results. Such order allocation problems are solved by techniques like linear programming, non-linear programming, mixed-integer programming, and artificial intelligence technique.

Linear Programming, which plays a vital role in the field of Operations Research, is a discipline that deals with the optimization and control of systems. The term "Programming" used in linear programming is a synonym for the word "Optimization". Linear programming models, in general, either maximize the output for a given set of inputs or minimize the input for achieving the required or desired output. The above maximization or minimization concept is applicable to any problem depending on the environment in which it occurs including the constraints, restrictions, and data from the decision-maker.

Linear programming problems are basically mathematical models which attempt to create a model for the given real-life situation under study. A real-life situation is converted into a mathematical model by using variables and parameters which are basically numbers. The difference between variables and parameters is that the latter are numbers known to the decision-maker and fixed in the model while the former are those numbers which shall be determined in the process. In certain literature, variables are also termed decision variables.

Linear programming problems consist of two main components called Constraints and Objectives. Objectives are framed by the decision-maker in line with their ultimate requirement or the purpose. An objective can be considered as the synonym for the Goal or Aim of the system. There can be single or multiple objectives in a single linear programming model. Some of the examples of objectives in a model are maximizing the profit of an organization, minimizing the products produced with defects, maximizing the total value of purchasing, maximizing the sales, etc. It is observed that since the linear programming models are linked to the concept of optimization, the objectives are always accompanied by Maxima or Minima function.

Constraints in a linear programming model cannot be violated or ignored. They exist in the environment or the scenario. They are compulsorily present in the system. A few examples of constraints used in various literature are capacity of a supplier, lead time of the product, defects in a product, product cycle time, financial constraints, resource limitations, contractual agreements, etc. It is generally advised to write down the constraint of the model as a sentence in English so that it can be thoroughly understood for conversion to a mathematical structure. A basic mathematical structure of a constraint can be represented as LHS R RHS, where the LHS means the actual value, R depicts a mathematical relationship $\{\leq, =, \geq\}$, and RHS means the Limit with which the LHS value has to be compared. For example, considering the budget as one of the constraints in a model, LHS will denote the money spent, RHS will denote the budget allocated, and R should be $\leq$ since we are always supposed to or expected to spend less than or equal to the allocated budget.

12.4 CASE STUDY

Outsourcing in Bharat Heavy Electricals Limited (BHEL) started way back in the 1960s as "Ancillary Development" in order to support and develop the people and surroundings near BHEL. Outsourcing department in BHEL, Trichy is responsible for outsourcing various products like pressure parts fabrication, attachment fabrication for shop assemblies, part processing fabrication for shop assemblies, structures, columns, ceiling girders, ducts, oil systems, feeders, hangers and suspensions and other miscellaneous fabrication, and punching and shearing jobs. The department functions and operates with 250 vendors located in and around Trichy. Evaluation of these vendors' performance becomes very crucial for the smooth and transparent functioning of the supply chain.

Mixed-Integer Linear Programming (MILP) model is developed for optimum order allocation and solved using MATLAB software. The model is explained using illustrative examples from the industry.

Order Allocation process in the firm is studied and analysed in order to derive a linear programming model. This process in the firm begins with the finalization of the contract with the vendors. Inputs for the finalization of the contract will be provided by the planning section of the firm to the contract section. The following points will be clearly spelt out in the inputs given by the planning section. Work content and scope of work to be carried out by the vendors, quantum of work in metric tons, list of vendors to be contacted for executing the job, validity of the contract, delivery period, engineering drawings to be followed, reference-quality documents and other techno-commercial terms and conditions.

Based on the inputs, the contract section will float a tender to the vendors. As per the techno-commercial tender terms and conditions, Load will be allocated to only those vendors who qualify in certain criteria like eligibility for the execution of the job, ranking of the vendor in the tender, availability of necessary customer approval to execute the job of that particular customer, establishment of the required bank guarantee by the vendor, good Vendor Performance Rating (VPR) score for the vendor, no backlog orders pending beyond 3 months for the vendor, sparable capacity for the vendor to execute the job.

The essence of any tendering would be to identify the least bidder for executing the job for the firm. Likewise, the vendors would be sorted based on their bids and provided with "Tender Ranking". Least bidder would be called L1 bidder and the next least would be called L2 and so on. In case of a tie, negotiation with vendors or lot system would be adopted by the firm to declare the ranking of the vendor. At the end of the tendering process, vendors would be given respective tender rankings L1, L2, etc., based on their bids.

After the finalization of the abovesaid tender conditions and signing of the contract by the vendor, the planning section will proceed with order allocation as per tender ranking and other mentioned conditions. Ordering by the planning section will depend on the load available in hand. In general, order allocation happens in a frequency of twice a week.

Linear Programming (LP) models maximize or minimize single or multiple linear objective functions subjected to certain constraints. The MILP model adds one additional condition to the LP model that at least one of the variables should definitely be an integer value, while other variables are allowed to be non-integers. The MILP model consists of parameters, input data, decision variables, objective function, and constraints.

With reference to the case study firm, the MILP model formulation is briefed below.

Indices
i = Vendors; c = Customers

Parameters
d_c = Demand of the customer c
q_{ic} = Qualification of vendor i to supply customer c

$$= \begin{cases} 1, \text{ if vendor i is qualified to supply customer c} \\ 0, \text{ otherwise} \end{cases}$$

nt_{ic} $= \begin{cases} 1, \text{ if vendor i is qualified to supply customer c} \\ 0, \text{ otherwise} \end{cases}$

POB_i $=$ PO (Purchase Order) Block of vendor i

$$= \begin{cases} 1, \text{ if not blocked} \\ 0, \text{ if blocked} \end{cases}$$

BG_i (Bank Gaurantee)$=1$, if Available

$$= \begin{cases} 0, \text{ if Not Available} \end{cases}$$

VPR_i $= \begin{cases} 1, \text{ if VPR of vendor i is} > 75 \end{cases}$

$$= \quad 0, \text{ otherwise}$$

Cap_i $=$ Capacity of vendor i

$Rank_{ic}$ $=$ Rank of vendor i for customer c

Decision Variables

X_{ic} $=$ Allocation in MT (Metric Tonne) to vendor i for customer c

δ_{ic} $= \begin{cases} 1, \text{ if vendor i is supplying to customer c} \\ 0, \text{ otherwise} \end{cases}$

Objective

Minimize $\Sigma_i \Sigma_c Rank_{ic} * \delta_{ic}$

Subject to Constraints

$\Sigma_i X_{ic} (q_{ic} * POB_i * BG_i * nt_{ic} * VPR_i) = d_c, \forall_c$	*(Demand Constraint)*
$\Sigma_c X_{ic} (q_{ic} * POB_i * BG_i * nt_{ic} * VPR_i) \leq Cap_i, \forall_i$	*(Capacity Constraint)*
$X_{ic} \leq Cap_i * \delta_{ic}, \forall_{i, c}$	*(Capacity Constraint)*
$X_{ic} \geq 0$	*(Non-negativity Constraint)*
$\delta_{ic} \, \varepsilon \, \{0, 1\}$	*(Vendor Load Allocation Constraint)*

As per the above model, the input parameters such as Demand, Qualification, Customer Approval, Block Status, VPR Score, BG Details, Capacity and Rank of the vendor in the tender will be available to the decision-maker of the firm. The objective of the model is to ensure that the order allocation is done as per tender ranking, i.e., load to be allocated from L1 bidder and so on. Hence, Minimization of Rank is chosen as the objective function. Constraints such as Demand, Vendor Capacity, Tender Minimum Load, and Non-negativity are expressed in the form of equations using the parameters available. The above MILP model is solved in MATLAB software. The model is explained by the following illustration chosen from the firm.

Illustration 1:

Table 12.1 indicates the demand data of the case study firm. The firm has a demand of 200 MT which is split as 4 WBS (Work Breakdown Strucure), where in WBS 1 belongs to NTPC customer which requires load allocation to be done only on an NTPC (National Thermal Power Corporation Limited)-approved vendor. The Rate Schedules (RS 1 and RS 2) in which each of the WBS falls are also indicated in the same table (Table 12.1).

Table 12.1 Demand data

Customer	NTPC Indicator	Rate Schedule Number	Rate Schedule Weight
WBS 1	Y	RS 1	60
WBS 2	N	RS 1	60
WBS 3	N	RS 2	35
WBS 4	N	RS 2	45
			200

Table 12.2 Vendor input parameters

Rate Schedule Number	Vendor	Rank	Vendor NTPC Indicator	Purchase Order Block	Bank Guarantee Indicator	VPR	Vendor Capacity
RS 1	A	L 1	N	N	Y	75	150
RS 1	B	L 2	N	Y	Y	65	120
RS 1	C	L 3	**Y**	N	Y	80	180
RS 2	W	L 1	N	N	Y	80	70
RS 2	X	L 2	Y	Y	Y	60	100
RS 2	Y	L 3	N	N	Y	85	100
RS 2	Z	L 4	N	Y	Y	65	80

Vendor-related parameters are indicated in Table 12.2. Parameters required for the MILP model like NTPC Indicator (Y=1, N=0), PO Block Status (Y=0, N=1), BG Indicator (Y=1, N=0), Vendor Capacity in MT, and Vendor Performance Rating are updated in the input table.

Based on the input parameters and objective function, the MILP model is solved using the MATLAB software and the results are summarized in Table 12.3. Table 12.4 indicates the iterative solution along with the reason for optimized load allocation as per the model. MATLAB output for the above illustration is shown in Figure 12.1.

Table 12.3 Order allocation summary

Customer	NTPC Indicator	Rate Schedule Number	Rate Schedule Weight	Vendor Identification
WBS 1	Y	RS 1	60	**C**
WBS 2	N	RS 1	60	**A**
WBS 3	N	RS 2	35	**W**
WBS 4	N	RS 2	45	**Y**
			200	

Table 12.4 Iterative solution with reasons

Customer	NTPC Indicator	Rate Schedule Number	Rate Schedule Weight	Vendor	Rank	Vendor NTPC Indicator	Purchase Order Block	Bank Guarantee Indicator	VPR	Vendor Capacity	Load Indicator	Reason
WBS 1	Y	RS 1	60	A	L1	N	N	Y	75	150	N	A is not an NTPC vendor, but WBS 1 requires an NTPC vendor.
				B	L2	N	Y	Y	65	120	N	B is not an NTPC vendor, but WBS 1 requires an NTPC vendor.
				C	L3	Y	N	Y	80	180	Y	C is NTPC-approved, No PO Block, BG Yes, VPR>70, Cap>Dem
WBS 2	N	RS 1	60	A	L1	N	N	Y	75	150	Y	WBS 4 does not require an NTPC vendor, No PO Block, BG Yes, VPR>70, Cap>Dem
				B	L2	N	Y	Y	65	120	NA	No demand
				C	L3	Y	N	Y	80	180	NA	No demand
WBS 3	N	RS 2	35	W	L1	N	N	Y	70	150	Y	WBS 2 does not require an NTPC vendor, No PO Block, BG Yes, VPR>70, Cap>Dem
				X	L2	Y	Y	Y	60	100	NA	No demand
				Y	L3	N	N	Y	85	100	NA	No demand
				Z	L4	N	Y	Y	65	80	NA	No demand
WBS 4	N	RS 2	45	W	L1	N	N	Y	35	150	N	Already Idn WBS 2 - 35 MT; revised capacity is 35 MT only.
				X	L2	Y	Y	Y	60	100	N	PO Block, VPR<70
				Y	L3	N	N	Y	85	100	Y	WBS 2 does not require an NTPC vendor, No PO Block, BG Yes, VPR>70, Cap>Dem
				Z	L4	N	Y	Y	65	80	NA	No demand

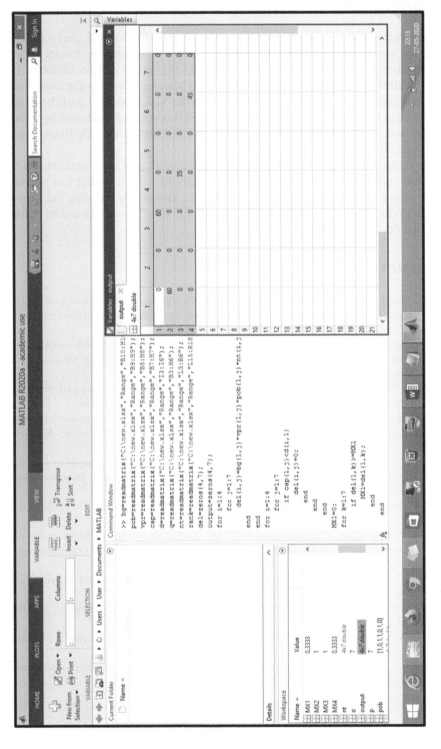

Figure 12.1 MATLAB coding and output for MILP model

12.5 CONCLUSION

This research paper discusses the Order Allocation process in a heavy industry in India. The aim of the research is to develop a linear programming model in order to ensure optimum order allocation on the selected suppliers of a heavy industry. It elaborates on all the critical factors involved in order allocation in the case study industry. A mathematical algorithm was developed using the MILP concept and the model has been tested in the MATLAB program. The model was verified and tested by using data from the case study. Model Output in MATLAB and the Case Study Illustration have been discussed in this chapter.

This study has resulted in developing a transparent and systematic model for the case study industry in order allocation on its suppliers. This model has resulted not only in saving time, energy, and resources for the case study industry but also maintain a fair relationship with its suppliers by utilizing this transparent model in its order allocation decision-making.

REFERENCES

Faez, F., Ghodsypour, S. H., and O'Brien, C. (2009). Vendor selection and order allocation using an integrated fuzzy case-based reasoning and mathematical programming model. *Int. J. Prod. Econ.*, 121(2), 395–408.

Ghodsypour, S. H., and O'Brien, C. (1998). A decision support system for supplier selection using an integrated analytic hierarchy process and linear programming. *Int. J. Prod. Econ.*, 56–57, 199–212.

Jayaraman, V., Srivastava, R., and Benton, W. C. (1999). Supplier selection and order quantity allocation: A comprehensive model. *J. Supply Chain Manag.*, 35(1), 50–58.

Liu, W., Liu, C., and Ge, M. (2013). An order allocation model for the two-echelon logistics service supply chain based on cumulative prospect theory. *J. Purch. Supply Manag.*, 19(1), 39–48.

Nazari-Shirkouhi, S., Shakouri, H., Javadi, B., and Keramati, A. (2013). Supplier selection and order allocation problem using a two-phase fuzzy multi-objective linear programming. *Appl. Math. Model.*, 37(22), 9308–9323.

Pan, A. C. (1989). Allocation of order quantity among suppliers. *J. Purch. Mater. Manag.*, 25(3), 36–39.

Razmi, J., Songhori, M. J., and Khakbaz, M. H. (2009). An integrated fuzzy group decision making/fuzzy linear programming (FGDMLP) framework for supplier evaluation and order allocation. *Int. J. Adv. Manuf. Technol.*, 43(5–6), 590–607.

Rosenthal, E. C., Zydiak, J. L., and Chaudhry, S. S. (1995). Vendor selection with bundling. *Decis. Sci.*, 26(1), 35–48.

Turner, I. (1988). An independent system for the evaluation of contract tenders. *J. Oper. Res. Soc.*, 39(6), 551–561.

Weber, C. A., Current, J. R., and Benton, W. C. (1991). Vendor selection criteria and methods. *Eur. J. Oper. Res.*, 50(1), 2–18.

Zeydan, M., Çolpan, C., and Çobanoğlu, C. (2011). A combined methodology for supplier selection and performance evaluation. *Expert Syst. Appl.*, 38(3), 2741–2751.

Chapter 13

Formulation of an optimal ordering policy with quadratic demand

Weibull distribution deterioration and partial backlogging

Prachi Swain, Chittaranjan Mallick, and Trailokyanath Singh

CONTENTS

13.1 INTRODUCTION

The concept of deterioration is not new. It is one of the categories of inventoried goods. Items like alcohol, gasoline, human blood, vegetable, radioactive substances, etc., having no self-life at all are called decaying items. Many researchers studied inventory problems and their solutions from time to time. Ghare and Schrader (1963) first studied the decaying inventory model in which commodities deteriorate by the negative exponential function of time. Aggarwal (1978) established the model with constant demand, constant deterioration, and the average holding cost. The inventory model for the replenishment and no shortages is presented by Dave and Patel (1981). Covert and Philip (1973) formulated the model without shortages on the consideration of two-parameter Weibull distribution deterioration. The comprehensive literature review of the models with deterioration have been published in the review papers of Raafat (1991) and Goyal and Giri (2001).

The concept of constant demand rate is not always suitable for the development of models on inventory items. Hollier and Mak (1983) studied the replenishment policies with different replenishment intervals. Wee (1995) and Jalan and Chaudhuri (1996) presented their model by considering rapid changes in demand patterns. Ghosh and Chaudhuri (2006) presented the model with the generalised demand as a quadratic function of time and constant deterioration.

DOI: 10.1201/9781003293576-13

As partial backlogging is a function of waiting time, the willingness of a customer to wait during this period is reciprocal to the length of the waiting time. It is applicable to fashionable items and high-tech products like supercomputers, laptops, android mobiles, etc. In this context, Chang and Dye (1999) developed the deteriorating models with time-varying demands and partial backlogging. Ouyang et al. (2005) studied the backlogging inventory model with exponentially declining demand in which the backlogging rate is reciprocal to the waiting time for the next replenishment. The three-parameter Weibull distributed inventory model with quadratic demand pattern is presented by Singh et al. (2021). Some models were developed with the consideration of quadratic holding cost. In this regard, Swain et al. (2021) formulated the model with both generalised demand pattern and deterioration.

The purpose of the present paper is to develop the optimal ordering policy where the demand is taken as a quadratic function of time, deterioration is taken as a two-parameter Weibull distribution, and partial backlogging is considered as the reciprocal of the length of waiting time for the next replenishment. An analytical solution is illustrated with the help of a numerical example. In the end, a sensitivity analysis of the parameters is made to validate.

13.2   ASSUMPTIONS AND NOTATIONS

The following assumptions and notations are useful for the presentation of the model:

i.   The system needs only one type of item and replenishment takes place at an infinite rate.

ii. $R(t) = \begin{cases} a + bt + ct^2, & I(t) > 0 \\ K_0, & I(t) \leq 0 \end{cases}$: Quadratic demand at any time t, where

$a, b,$ and c $(a \geq 0, b \neq 0,$ and $c \neq 0)$ and K_0 are positive constants. Here $a, b,$ and c are the initial, increased, and self-increased demand rates, respectively.

iii. $\theta_0(t) = \alpha_0 \beta t^{\beta-1}$: Two-parameter Weibull distribution deterioration rate, where $\alpha_0 (0 < \alpha_0 \ll 1)$ and $\beta(> 0)$ are taken as the scale and the shape parameters, respectively.

iv. A small part of demand for the shortages is considered as backlogged. The waiting time is inversely proportional to backlogging rate. Then, at that point, the multiplying rate will be more modest. Customers who would like to agree to backlogging at any time $(T - t)$ is diminishing with waiting time for the subsequent replenishment. For negative stock, $B_0(t) = \dfrac{1}{1 + \delta_0(T - t)}$ is the backlogging rate in the span $t_B \leq t \leq T$, where δ_0 is positive.

The notations are as follows:

$C_1, C_2, C_3, C_4,$ and C_5: Holding cost, $ per unit per unit time, cost of the item, $ per unit, ordering cost, $ per order, shortage cost, $ per unit per unit time and opportunity cost due to lost sales, $ per unit, respectively.

t_B and T: Starting time of the shortage and span of each cycle, respectively.

W_{max}, S_{max}, and Q_0: Highest level for each ordering cycle, maximum quantity of backlogged demand, and Economic Order Quantity (EOQ) for each ordering cycle, respectively.

$I(t)$: Inventory level at any time t.

13.3 MATHEMATICAL FORMULATION

At the initial time, the inventory level reaches its maximum W_{max} units of item when time $t = 0$. During the period $[0, t_B]$, the inventory is reduced due to both demand and Weibull distribution deterioration. Further, the level of inventory accomplishes zero and all of the demand is partially backlogged in the interval $[t_B, T]$.

The inventory can be represented by the following differential equation:

$$\frac{dI(t)}{dt} + \theta_0(t)I(t) = -R(t), \quad 0 \le t \le t_B \tag{13.1}$$

where $\theta_0(t) = \alpha_0 \beta t^{\beta-1}$ and $R(t) = a + bt + ct^2$.

Using the boundary condition $I(t_B) = 0$, the solution of Eq. (13.1) is given by

$$I(t) = \left[a\left(t_B + \frac{\alpha_0 t_B^{\beta+1}}{\beta+1} \right) + b\left(\frac{t_B^2}{2} + \frac{\alpha_0 t_B^{\beta+2}}{\beta+2} \right) + c\left(\frac{t_B^3}{3} + \frac{\alpha_0 t_B^{\beta+3}}{\beta+3} \right) \right.$$
$$\left. - \left\{ a\left(t + \frac{\alpha_0 t^{\beta+1}}{\beta+1} \right) + b\left(\frac{t^2}{2} + \frac{\alpha_0 t^{\beta+2}}{\beta+2} \right) + c\left(\frac{t^3}{3} + \frac{\alpha_0 t^{\beta+3}}{\beta+3} \right) \right\} \right] e^{-\alpha_0 t^\beta}, \quad 0 \le t \le t_B \tag{13.2}$$

(by ignoring the terms of α_0^2 as $0 < \alpha_0 \ll 1$).

Using the boundary condition $I(0) = W_{max}$ in Eq. (13.2), the maximum level of inventory for each cycle becomes

$$I(0) = W_{max} I(0) = W_{max} = a\left(t_B + \frac{\alpha_0 t_B^{\beta+1}}{\beta+1} \right) + b\left(\frac{t_B^2}{2} + \frac{\alpha_0 t_B^{\beta+2}}{\beta+2} \right) + c\left(\frac{t_B^3}{3} + \frac{\alpha_0 t_B^{\beta+3}}{\beta+3} \right) \tag{13.3}$$

During the shortage interval $[t_B, T]$, the demand rate is considered as partially backlogged at the fraction $\dfrac{1}{1 + \delta_0(T-t)}$.

Therefore, the representation of the inventory system is given by

$$\frac{dI(t)}{dt} = -\frac{K_0}{1 + \delta_0(T-t)}, \quad t_B \le t \le T \tag{13.4}$$

Using the boundary condition $I(t_B) = 0$, the solution of Eq. (13.4) is

$$I(t) = \frac{K_0}{\delta_0} \left[\ln\{1 + \delta_0 (T - t)\} - \ln\{1 + \delta_0 (T - t_1)\} \right], \quad t_B \leq t \leq T \tag{13.5}$$

Maximum backlogged demand per cycle is computed by setting $t = T$ in Eq. (13.5). Therefore,

$$S_{max} = -I(t) = \frac{K_0}{\delta_0} \ln\left[1 + \delta_0 (T - t_B) \right] \tag{13.6}$$

and the EOQ per cycle is given by

$$Q_0 = W_{max} + S_{max}$$

$$= a\left(t_B + \frac{\alpha t_B^{\beta+1}}{\beta+1} \right) + b\left(\frac{t_B^2}{2} + \frac{\alpha_0 t_B^{\beta+2}}{\beta+2} \right) + c\left(\frac{t_B^3}{3} + \frac{\alpha_0 t_B^{\beta+3}}{\beta+3} \right) + \frac{K_0}{\delta_0} \ln\left[1 + \delta_0 (T - t_B) \right] \tag{13.7}$$

Holding cost (I_{HC}) per cycle:

$$I_{HC} = C_1 \int_0^{t_B} I(t)dt$$

$$= C_1 \left[a\left(\frac{t_B^2}{2} + \frac{\alpha_0 \beta t_B^{\beta+2}}{(\beta+1)(\beta+2)} \right) + b\left(\frac{t_B^3}{3} + \frac{\alpha_0 \beta t_B^{\beta+3}}{(\beta+1)(\beta+3)} \right) + c\left(\frac{t_B^4}{4} + \frac{\alpha_0 \beta t_B^{\beta+4}}{(\beta+1)(\beta+4)} \right) \right] \tag{13.8}$$

(by ignoring the terms of α_0^2 as $0 < \alpha_0 \ll 1$)
Deterioration cost (I_{DC}) per cycle:

$$I_{DC} = C_2 \left[W_{max} - \int_0^{t_B} D(t)dt \right] = \alpha_0 C_2 \left(\frac{a t_B^{\beta+1}}{\beta+1} + \frac{b t_B^{\beta+2}}{\beta+2} + \frac{c t_B^{\beta+3}}{\beta+3} \right) \tag{13.9}$$

Shortage cost (I_{SC}) per cycle:

$$I_{SC} = C_4 \left[-\int_0^{t_B} I(t)dt \right] = C_4 K_0 \left[\frac{T - t_B}{\delta_0} - \frac{1}{\delta_0^2} \ln\{1 + \delta_0 (T - t_B)\} \right] \tag{13.10}$$

Opportunity cost (I_{OC}) due to lost sales per cycle:

$$I_{OC} = C_5 \int_{t_B}^{T} K_0 \left[1 - \frac{1}{1 + \delta_0 (T - t)} \right] dt = C_5 K_0 \left[T - t_B - \frac{1}{\delta_0} \ln\{1 + \delta_0 (T - t_B)\} \right] \tag{13.11}$$

Therefore, $AC(t_B, T)$, the average total cost per unit time per cycle is

$$AC(t_B, T) = \frac{1}{T}\left[I_{HC} + I_{DC} + C_3 + I_{SC} + I_{OC}\right]$$

$$= \frac{1}{T}\left[C_1\left[a\left(\frac{t_B^2}{2} + \frac{\alpha_0 \beta t_B^{\beta+2}}{(\beta+1)(\beta+2)}\right) + b\left(\frac{t_B^3}{3} + \frac{\alpha_0 \beta t_B^{\beta+3}}{(\beta+1)(\beta+3)}\right) + c\left(\frac{t_B^4}{4} + \frac{\alpha_0 \beta t_B^{\beta+4}}{(\beta+1)(\beta+4)}\right)\right]\right.$$

$$+ \alpha_0 C_2\left(\frac{at_B^{\beta+1}}{\beta+1} + \frac{bt_B^{\beta+2}}{\beta+2} + \frac{ct_B^{\beta+3}}{\beta+3}\right) + C_3 + \frac{K_0(C_4 + \delta_0 C_5)}{\delta_0}$$

$$\left.\left[T - t_B - \frac{1}{\delta_0}\ln\{1 + \delta_0(T - t_B)\}\right]\right]$$

$$(13.12)$$

The values of t_B^* and T^* for the minimum average cost AC are obtained from the equations

$$\frac{\partial(AC)}{\partial t_B} = 0 \text{ and } \frac{\partial(AC)}{\partial T} = 0 \qquad (13.13)$$

provided that

$$\frac{\partial^2(AC)}{\partial t_B^2} > 0, \frac{\partial^2(AC)}{\partial T^2} > 0, \text{ and } \left[\frac{\partial^2(AC)}{\partial t_B^2} \times \frac{\partial^2(AC)}{\partial T^2} - \left(\frac{\partial^2(AC)}{\partial t_B \partial T}\right)^2\right] > 0 \qquad (13.14)$$

From Eq. (13.13), we have

$$\frac{\partial(AC)}{\partial t_B} = \frac{1}{T}\left[t_B\left(a + bt_B + ct_B^2\right)\left[C_1\left(1 + \frac{\alpha_0 t_B^{\beta+1}}{\beta+1}\right) + \alpha_0 C_2 t_B^{\beta-1}\right]\right.$$

$$\left. - \frac{K_0(C_4 + \delta_0 C_5)(T - t_B)}{1 + \delta_0(T - t_B)}\right] = 0 \qquad (13.15)$$

and

$$\frac{\partial(AC)}{\partial T} = \frac{1}{T}\left[\frac{K_0(C_4 + \delta_0 C_5)(T - t_B)}{1 + \delta_0(T - t_B)} - (AC)\right] = 0 \qquad (13.16)$$

Firstly, the optimal times t_B^* and T^* are computed from both Eqs. (13.15) and (13.16). Thereafter, the optimal order quantity Q_0^*, the optimal maximum inventory level W_{max}^*, and the minimum average total cost per unit time AC^* can be calculated by substituting the values of t_B^* and T^* in the Eqs. (13.7), (13.3), and (13.12), respectively.

13.4   NUMERICAL ILLUSTRATION

Example 1: For the inventory system, the following data is considered with their proper units: $a = 4$, $b = 3$, $c = 2$, $\delta_0 = 2$, $C_1 = 0.5$, $C_2 = 1.5$, $C_3 = 10$, $C_4 = 2.5$, $C_5 = 2$, $K_0 = 8$, $\alpha_0 = 0.01$, and $\beta = 2$. Solving Eqs. (13.14) and (13.15), we get $t_B^* = 1.44009$ unit time and $T^* = 1.72747$ unit time. Then, we get optimal order quantity $Q_0^* = 12.7754$ units, the optimal maximum inventory level $W_{max}^* = 10.959$ units and the minimum average total cost per unit time $AC^* = 9.48949$.

13.5   SENSITIVITY ANALYSIS

The sensitivity analysis is carried out by changing the parameters one at a time by +50%, +10%, −10%, and −50% and keeping fixed the remaining parameters illustrated in Example 1. The notable points are observed from Table 13.1.

i. t_B^* and T^* decrease while Q_0^* and AC^* increase in that of $a, b,$ and C_1. Here, $t_B^*, T^*,$ and AC^* are moderately sensitive and Q_0^* is highly sensitive to change in $a, b,$ and C_1.

ii. $t_B^*, T^*,$ and Q_0^* decrease while AC^* increases in that of c and C_2. Here, t_B^* and T^* are moderately sensitive while Q_0^* and AC^* are less sensitive to change in c and C_2.

iii. t_B^* and AC^* increase while T^* and Q_0^* decrease in that of $\delta_0, C_4,$ and C_5. Here, all $t_B^*, T^*, Q_0^*,$ and AC^* are moderately sensitive to change in $\delta_0, C_4,$ and C_5.

iv. $t_B^*, T^*, Q_0^*,$ and AC^* increase with increase in that of C_3. Here, all $t_B^*, T^*, Q_0^*,$ and AC^* are highly sensitive to change in C_3.

v. $t_B^*, Q_0^*,$ and AC^* increase while T^* decreases with increase in that of K_0. Here, all $t_B^*, T^*, Q_0^*,$ and AC^* are moderately sensitive to change in K_0.

vi. $t_B^*, T^*,$ and Q_0^* decrease while AC^* increases with increase in the value of α_0 and β. Here, all $t_B^*, T^*, Q_0^*,$ and AC^* are less sensitive to change in α_0 and β.

13.6   CONCLUSIONS

In the present paper, an inventory model for two-parameter Weibull distribution deteriorating items with quadratic demand and partial backlogging is considered. The inspiration for taking on the present model is the thought of the quadratic demand rate. This occurs because of the sped-up development or decrease in the interest rate. Sped-up development sought-after rate is found in high-tech products like android mobile, supercomputers, machines, and hardware, whereas speed decrease occurs on account of outdated computers, outdated aeroplanes and apparatuses, and so forth.

The proposed model can be reached out in a few ways. Initially, we might stretch out the quadratic demand to a more summed up request design that vacillates with time. We could extend the proposed two-parameter Weibull distributed model to a three-parameter Weibull distributed model. Besides, we could extend the deterministic model into stochastic model and quantity discount model.

Table 13.1 Sensitivity analysis

Parameters	% change in parameter	t_B^*	T^*	Q_0^*	AC^*
a	+50	1.37598	1.69304	14.8958	10.0892
	+10	1.42701	1.72032	13.2145	9.61293
	−10	1.4533	1.73475	12.3287	9.36425
	−50	1.50743	1.76522	10.4629	8.84489
b	+50	1.36086	1.66877	13.3007	9.9091
	+10	1.423	1.71463	12.8876	9.57816
	−10	1.45789	1.74093	12.6591	9.39819
	−50	1.53718	1.80204	12.1464	9.0034
c	+50	1.36156	1.66326	12.7271	9.78436
	+10	1.42239	1.71282	12.7613	9.55311
	−10	1.45905	1.74327	12.7926	9.42309
	−50	1.55192	1.8222	12.9062	9.12299
δ_0	+50	1.44971	1.69558	12.5553	9.62191
	+10	1.44234	1.72026	12.7227	9.52031
	−10	1.43763	1.73518	12.834	9.45589
	−50	1.42496	1.77246	13.1535	9.28383
C_1	+50	1.25556	1.6024	10.8715	10.6489
	+10	1.39586	1.69591	12.2854	9.75115
	−10	1.48955	1.76375	13.3488	9.20847
	−50	1.77384	1.98839	17.1832	7.80664
C_2	+50	1.42923	1.71859	12.6478	9.53104
	+10	1.43788	1.72566	12.7494	9.49788
	−10	1.44232	1.72929	12.8018	9.48106
	−50	1.45141	1.73677	12.9098	9.44692
C_3	+50	1.61708	2.05477	15.8769	12.136
	+10	1.4807	1.79611	13.4399	10.0571
	−10	1.39598	1.65614	12.0823	8.89841
	−50	1.1658	1.32293	8.89481	6.21692
C_4	+50	1.45049	1.6759	12.5804	9.63268
	+10	1.44253	1.71497	12.7299	9.52295
	−10	1.43742	1.74143	12.8251	9.45297
	−50	1.42365	1.8185	13.0793	9.26612
C_5	+50	1.45507	1.65455	12.4936	9.69622
	+10	1.44389	1.70809	12.7043	9.5417
	−10	1.43569	1.75063	12.8571	9.42941
	−50	1.4083	1.91512	13.3589	9.0609
K_0	+50	1.46096	1.62827	12.9585	9.77835
	+10	1.44602	1.69751	12.8274	9.571
	−10	1.43258	1.76753	12.7097	9.38704
	−50	1.36171	2.2925	12.0925	8.45706
α_0	+50	1.42568	1.71553	12.6521	9.54126
	+10	1.43714	1.72502	12.7501	9.49998
	−10	1.44308	1.72995	12.8012	9.47892
	−50	1.4554	1.74023	12.908	9.4359
β	+50	1.42439	1.71261	12.5924	9.50697
	+10	1.43721	1.72474	12.7415	9.49254
	−10	1.44284	1.73009	12.8081	9.48668
	−50	1.45279	1.73963	12.93	9.47821

REFERENCES

Aggarwal, S. P. (1978). A note on an order-level inventory model for a system with constant rate of deterioration. *Opsearch*, 15, 184–187.

Chang, H. J., & Dye, C. Y. (1999). An EOQ model for deteriorating items with time varying demand and partial backlogging. *Journal of the Operational Research Society*, 50, 1176–1182.

Covert, R. P., & Philip, G. C. (1973). An EOQ model for items with Weibull distribution deterioration. *AIIE Transaction*, 5, 323–326.

Dave, U., & Patel, L. K. (1981). (T, Si) Policy inventory model for deteriorating items with time proportional demand. *Journal of the Operational Research Society*, 32, 137–142.

Ghare, P. M., & Schrader, G. P. (1963). A model for exponentially decaying inventories. *Journal of Industrial Engineering*, 14, 238–243.

Goyal, S. K., & Giri, B. C. (2001). Recent trends in modeling of deteriorating inventory. *European Journal of Operational Research*, 134, 1–16.

Hollier, R. H., & Mak, K. L. (1983). Inventory replenishment policies for deteriorating items in a declining market. *International Journal of Production Research*, 21, 813–826.

Jalan, A., Giri, R., & Chaudhury, K. (1996). EOQ model for items with Weibull distribution deterioration shortages and trended demand. *International Journal of Production Economics*, 113, 852–861.

Ghosh, S. K., & Chaudhuri, K. S. (2006). An EOQ model with a quadratic demand, time proportional deterioration and shortages in all cycles. *International Journal of Systems Sciences*, 37(10), 663–672.

Ouyang, L. Y., Wu, K. S., & Cheng, M. C. (2005). An inventory model for deteriorating items with exponential declining demand and partial backlogging. *Yugoslav Journal of Operations Research*, 15(2), 277–288.

Raafat, F. (1991). Survey of literature on continuously deteriorating inventory model. *Journal of the Operational Research Society*, 42, 27–37.

Singh, T., Pattanayak, H., Nayak, A. K., & Sethy, N. N. (2021). An optimal policy with three-parameter Weibull distribution deterioration, quadratic demand, and salvage value under partial backlogging. *International Journal of Rough Sets and Data Analysis*, 5(1), 79–98.

Swain, P., Mallick, C., Singh, T., Mishra, P. J., & Pattanayak, H. (2021). Formulation of an optimal ordering policy with generalised time-dependent demand, quadratic holding cost and partial backlogging. *Journal of Information and Optimization Sciences*, 42(5), 1163–1179.

Wee, H. M. (1995). A deterministic lot-size inventory model for deteriorating items with shortages and a declining market. *Computers and Operations Research*, 22, 345–356.

Chapter 14

An optimal replenishment policy with exponential declining demand

Weibull distribution deterioration and partial backlogging

Sephali Mohanty, Trailokyanath Singh, and Sudhansu Sekhar Routary

CONTENTS

14.1 INTRODUCTION

In recent decades, many researches on the replenishment policy for deteriorating items such as food items, drugs, volatile substances, radioactive substances, pharmaceuticals and others have been widely studied. The term "deterioration" is defined as damage, spoilage or decay such that the items cannot be used for their original purpose. The earliest researcher Ghare and Schrader (1963) developed a simple EOQ (Economic Order Quantity) inventory model by assuming a constant deterioration rate. Next, Covert and Philip (1973) modified Ghare and Schraders' model by using a two-parameter Weibull distribution for time-dependent deterioration rate. Dave and Patel (1981) considered an inventory system by allowing time variability in one or more parameters. Detailed information regarding the development of deteriorated inventory model was provided in the review articles of Raafat (1991) and Goyal and Giri (2001). Singh et al. (2017) established the optimal replenishment policy with the help of the stock-dependent demand and the concept of permissible delay in payment.

For fashion goods and high-tech products with a short product life cycle, the waiting time of a customer for backlogging during shortage is diminishing with the length of the waiting time. To reflect this phenomenon, Chang and Dye (1999) developed an inventory model in a linear function of the waiting time. The age of the inventory has

DOI: 10.1201/9781003293576-14

a negative impact on demand due to the loss of consumer confidence in the quality of products like fashionable commodities, high-tech products, etc., and physical loss of materials. Hollier and Mak (1983) assumed the optimal replenishment policies on the consideration of exponentially declining demand with different replenishment intervals. Later, Wee (1995) studied the inventory model for deteriorating items with the exponential declining demand. During the shortage period, backlogging rate acts as a variable and is dependent on the waiting time for the next replenishment. Recently, Ouyang et al. (2005) established the partially backlogged model for deteriorating items with exponential declining demand on the assumption of the constant deterioration rate. However, the Weibull distribution deterioration is used to represent the time to deterioration. Therefore, the inventory model for deteriorating items with exponential declining demand, Weibull distribution and partial backlogging is more realistic than exponential declining demand and partial backlogging. Singh et al. (2021) derived an ordering policy varying with the three-parameter Weibull deterioration, the time-dependent quadratic demand, partial backlogging and salvage value. Swain et al. (2021) formulated the replenishment policy with generalised demand, quadratic holding cost and partial backlogging.

The motivation behind developing a replenishment model in the present paper is to develop a generalised inventory model which includes shortages, two-parameter Weibull distribution deterioration and exponentially declining demand. The backlogging rate is reciprocal to the waiting time for the next replenishment. The optimal order quantity has been derived by minimising the total relevant cost. A brief analysis of the total relevant cost involved is illustrated by a numerical example. Finally, sensitivity analysis of all parameters is carried out.

14.2   ASSUMPTIONS AND NOTATIONS

The model in this work is presented on the basis of the following assumptions and notations.

Assumptions

i.   The inventory system considers only one type of items.
ii. Replenishment rate is infinite.
iii. $\theta_0(t) = \alpha_0 \beta t^{\beta-1}$, where $0 < \alpha_0 \ll 1$ and $\beta > 0$, is time-dependent deterioration rate and is two-parameter Weibull distribution deterioration.
iv. Neither replacement nor repairing of deteriorated units will take place during the period under consideration.
v. $R(t) = \begin{cases} Ke^{-\gamma t}, & I(t) \geq 0 \\ D_0, & I(t) < 0 \end{cases}$: Demand rate, where $K(> 0)$ is initial demand and $\gamma(0 < \gamma \ll 1)$ is a constant governing the declining rate of the demand.
vi. During the shortage period, the backlogging rate acts as a variable and depends on the length of the waiting time for the next replenishment. The waiting time is reciprocal to the backlogging rate. In this situation, the backlogging rate is $\dfrac{1}{1+\delta_0(T-t)}$ when inventory is negative. The backlogging parameter δ_0 is a positive constant, $t_s \leq t \leq T$.

Notations

C_1, C_2, C_3, C_4 and C_5: Holding cost, $ per unit per unit time, cost of the inventory item, $ per unit, ordering cost of inventory, $ per order, shortage cost, $ per unit per unit time and opportunity cost due to lost sales, $ per unit, respectively.

t_s and T: Time at which shortages start and length of each ordering cycle, respectively.

W_{max}, S_{max} and Q_0: Maximum inventory level for each ordering cycle, maximum amount of demand backlogged for each ordering cycle and EOQ for each ordering cycle, respectively.

$I(t)$: Inventory level at time t.

14.3 MODEL DEVELOPMENT AND ANALYSIS

At the start of the inventory model, the maximum level of inventory is W_{max} units of items at time $t = 0$. During the period $[0, t_s]$, the inventory level reduces due to demand and deterioration. At time t_s, the inventory level achieves zero, then shortage is allowed to occur and partially backlogged during the time interval $[t_s, T]$.

The inventory system can be represented by the following differential equations:

$$\frac{dI(t)}{dt} + \theta_0(t)I(t) = -Ke^{-\gamma t}, \quad 0 \le t \le t_s \tag{14.1}$$

where $\theta_0(t) = \alpha_0 \beta t^{\beta-1}$.

Using the boundary condition $I(t_s) = 0$, the solution of Eq. (14.1) is given by

$$I(t) = K\left[t_s + \frac{\gamma t_s^2}{2} + \frac{\alpha_0 t_s^{\beta+1}}{\beta+1} - \left(t + \frac{\gamma t^2}{2} + \frac{\alpha_0 t^{\beta+1}}{\beta+1}\right)\right]e^{-\alpha_0 t^\beta}, \quad 0 \le t \le t_s \tag{14.2}$$

(ignoring the terms of higher power of α_0 and γ as $0 < \alpha_0 \ll 1$ and $0 < \gamma \ll 1$)

Using the boundary condition $I(0) = W_{max}$, the maximum inventory level for each cycle is given by

$$I(0) = W_{max} = K\left[t_s + \frac{\gamma t_s^2}{2} + \frac{\alpha_0 t_s^{\beta+1}}{\beta+1} - \left(t + \frac{\gamma t^2}{2} + \frac{\alpha_0 t^{\beta+1}}{\beta+1}\right)\right] \tag{14.3}$$

During the shortage time interval $[t_s, T]$, the demand at any time t is partially back-logged at the fraction $\dfrac{1}{1+\delta_0(T-t)}$. Therefore, the inventory system is represented by the differential equation:

$$\frac{dI(t)}{dt} = -\frac{D_0}{1+\delta_0(T-t)}, \quad t_s \le t \le T \tag{14.4}$$

Using the boundary condition $I(t_s) = 0$, the solution of Eq. (14.4) is given by

$$I(t) = \frac{D_0}{\delta_0}\left[\ln\{1+\delta_0(T-t)\} - \ln\{1+\delta_0(T-t_s)\}\right], \quad t_s \le t \le T \tag{14.5}$$

The maximum amount of backlogged demand per cycle is obtained from Eq. (14.5). Therefore,

$$S_{max} = -I(T) = \frac{D_0}{\delta_0}\ln\left[1+\delta_0\left(T-t_s\right)\right] \qquad (14.6)$$

Hence, the EOQ per cycle is

$$Q_0 = W_{max} + S_{max} = K\left(t_s + \frac{\gamma t_s^2}{2} + \frac{\alpha_0 t_s^{\beta+1}}{\beta+1}\right) + \frac{D_0}{\delta}\ln\left[1+\delta_0\left(T-t_s\right)\right] \qquad (14.7)$$

The inventory holding cost per cycle is

$$HC = C_1\int_0^{t_s} I(t)dt = C_1 K\left[\frac{t_s^2}{2} + \frac{\gamma t_s^3}{3} + \frac{\alpha_0\beta t_s^{\beta+2}}{(\beta+1)(\beta+2)}\right] \qquad (14.8)$$

(ignoring the terms of the higher powers of α_0 and γ as $0<\alpha_0\ll 1$ and $0<\gamma\ll 1$). The deterioration cost per cycle is

$$DC = C_2\left[W - \int_0^{t_s} D(t)dt\right] = C_2 K\left[t_s + \frac{\gamma t_s^2}{2} + \frac{\alpha_0 t_s^{\beta+1}}{\beta+1} - \frac{1}{\gamma}\left(1-e^{-\gamma t_s}\right)\right] \qquad (14.9)$$

The shortage cost per cycle is

$$SC = C_4\left[-\int_{t_s}^{T} I(t)dt\right] = C_4 D_0\left[\frac{T-t_s}{\delta_0} - \frac{1}{\delta_0^2}\ln\left[1+\delta_0\left(T-t_s\right)\right]\right] \qquad (14.10)$$

The opportunity cost due to lost sales per cycle is

$$OC = C_5\int_{t_s}^{T}\left[D_0\left(1-\frac{1}{1+\delta_0\left(T-t\right)}\right)\right]dt = C_5 D_0\left[T-t_s - \frac{1}{\delta_0}\ln\left[1+\delta_0\left(T-t_s\right)\right]\right] \qquad (14.11)$$

Therefore, the average total cost per unit time per cycle, i.e.,

$$AT = AT\left(t_1,T\right) = \frac{[HC+DC+OC+SC+OC]}{T}$$

$$= \frac{1}{T}\left[C_1 K\left(\frac{t_s^2}{2} + \frac{\gamma t_s^3}{3} + \frac{\alpha_0\beta t_s^{\beta+2}}{(\beta+1)(\beta+2)}\right) + C_2 K\left[t_s + \frac{\gamma t_s^2}{2} + \frac{\alpha_0 t_s^{\beta+1}}{\beta+1} - \frac{1}{\gamma}\left(1-e^{-\gamma t_s}\right)\right] + C_3\right.$$

$$\left. + C_4 D_0\left[\frac{T-t_1}{\delta_0} - \frac{1}{\delta_0^2}\ln\left[1+\delta_0\left(T-t_s\right)\right]\right] + C_5 D_0\left[T-t_s - \frac{1}{\delta_0}\ln\left[1+\delta_0\left(T-t_s\right)\right]\right]\right]$$

$$(14.12)$$

The optimum values of t_s and T for the minimum cost AT are nothing but solutions to the equations

$$\frac{\partial(AT)}{\partial t_s} = 0 \text{ and } \frac{\partial(AT)}{\partial T} = 0 \tag{14.13}$$

provided that

$$\frac{\partial^2(AT)}{\partial t_s^2} > 0, \frac{\partial^2(AT)}{\partial T^2} > 0 \text{ and } \frac{\partial^2(AT)}{\partial t_s^2} \cdot \frac{\partial^2(AT)}{\partial T^2} - \left(\frac{\partial^2(AT)}{\partial t_s \, \partial T}\right)^2 > 0$$

From Eq. (14.13), we have

$$\frac{\partial(AT)}{\partial t_s} = \frac{1}{T}\left[C_1 K\left(t_s + \gamma t_s^2 + \frac{\alpha_0 \beta t_s^{\beta+1}}{\beta+1} \right) + C_2 K\left[1 + \gamma t_s + \alpha_0 t_s^\beta - e^{-\gamma t_s} \right] \right.$$
$$\left. - \frac{D_0 (C_4 + \delta_0 C_5)(T - t_s)}{1 + \delta_0 (T - t_s)} \right] = 0 \tag{14.14}$$

and

$$\frac{\partial(AT)}{\partial T} = \frac{1}{T}\left[\frac{D_0 (C_4 + \delta_0 C_5)(T - t_s)}{1 + \delta_0 (T - t_s)} - (AT) \right] = 0 \tag{14.15}$$

Now, t_s^* and T^* can be computed from both Eqs. (14.14) and (14.15). Thereafter, the optimal order quantity, the optimal inventory level and the optimal cost per unit time can be computed with values of t_s^* and T^*, from Eqs. (14.7), (14.3) and (14.12), respectively.

14.4 NUMERICAL EXAMPLE

Example 1: The following parameters are considered for solving equations of the model: $K = 12$, $\delta_0 = 2$, $\gamma = 0.03$, $C_1 = 0.5$, $C_2 = 1.5$, $C_3 = 10$, $C_4 = 2.5$, $C_5 = 2$, $D_0 = 8$, $\alpha_0 = 0.01$ and $\beta = 2$.

The optimal solutions obtained from the Eqs. (14.14) and (14.15) are found to be: the shortage period $t_s^* = 1.45072$ unit time and the length of ordering cycle $T^* = 1.82521$ unit time. Thereafter, the optimal values of the initial order quantity, maximum inventory level and the minimum average total cost per unit time are $Q_0^* = 20.1457$ units, $W_{max}^* = 17.9096$ units and $AT^* = 11.1341$, respectively.

14.5 SENSITIVITY ANALYSIS

The sensitivity analysis is considered from Example 1 by changing the value of each parameter by +50%, +10%,–10% and –50% with keeping other parameters fixed. The observed points are (Table 14.1):

Table 14.1 Performance of sensitivity analysis

Parameter	% change	t_s^*	T^*	Q^*	W^*	AT^*	% change in AT^*
K	+50	1.1425	1.63578	23.7525	21.0069	12.912	+15.97
	+10	1.37411	1.77357	20.975	18.6263	11.5469	+3.71
	−10	1.53854	1.88737	19.2478	17.1308	10.6847	−4.04
	−50	2.09474	2.33219	14.7015	13.1472	8.37163	−24.81
δ_0	+50	1.4712	1.80289	20.0132	18.1714	11.3054	+1.54
	+10	1.45545	1.82039	20.1132	17.97	11.1736	+0.35
	−10	1.44559	1.83022	20.1823	17.8441	11.0912	−0.39
	−50	1.41962	1.85244	20.3898	17.5126	10.8748	−2.33
γ	+50	1.38368	1.77881	19.5565	17.2271	11.477	+3.08
	+10	1.43658	1.81525	20.0214	17.7662	11.2049	+0.64
	−10	1.46527	1.83554	20.2737	18.0569	11.0621	−0.65
	−50	1.52797	1.88106	20.8255	18.6885	11.7613	−3.35
C_1	+50	1.18478	1.6595	17.2067	14.5365	12.6628	+13.73
	+10	1.3869	1.78225	19.4262	17.0957	11.4805	+3.11
	−10	1.52194	1.87512	20.9587	18.8212	10.7628	−3.33
	−50	1.9227	2.18541	25.7112	24.0221	8.95552	−19.57
C_2	+50	1.38276	1.77808	19.3855	17.043	11.4799	+3.11
	+10	1.43643	1.81513	19.9825	17.7271	11.2054	+0.64
	−10	1.46541	1.83565	20.314	18.0973	11.0617	−0.65
	−50	1.52841	1.88143	21.0403	18.9042	10.76	−3.36
C_3	+50	1.73738	2.28349	24.5546	21.6017	13.573	+21.9
	+10	1.51437	1.92143	21.1066	18.7242	11.668	+4.8
	−10	1.38301	1.72557	19.1336	17.0462	10.5708	−5.06
	−50	1.05372	1.27187	14.3396	12.8913	7.89792	−29.07
C_4	+50	1.47603	1.76467	20.056	18.2332	11.3459	+1.9
	+10	1.45668	1.81024	20.125	17.9857	11.1839	+0.45
	−10	1.4442	1.84216	20.1684	17.8263	11.0796	−0.49
	−50	1.41037	1.93957	20.2824	17.3947	10.7979	−3.02
C_5	+50	1.48711	1.74061	20.0154	18.3749	11.4388	+2.74
	+10	1.46001	1.80207	20.1133	18.0283	11.2117	+0.7
	−10	1.43997	1.85344	20.1829	17.7723	11.0443	−0.81
	−50	1.37252	2.06997	20.4061	16.9127	10.484	−5.84
D_0	+50	1.50128	1.71186	20.6652	18.5564	11.5578	+3.81
	+10	1.46519	1.78963	20.2949	18.0945	11.255	+1.09
	−10	1.43234	1.87442	19.9554	17.6749	10.9807	−1.38
	−50	1.25878	2.64269	18.1234	15.4704	9.54972	−14.23
α_0	+50	1.42903	1.80712	19.9436	17.691	11.1952	+0.55
	+10	1.44625	1.82147	20.1041	17.8646	11.1465	+0.11
	−10	1.45526	1.82901	20.188	17.9553	11.1223	−0.11
	−50	1.47416	1.84492	20.3642	18.1452	11.0705	−0.57
β	+50	1.42544	1.80079	19.8349	17.5949	11.1487	+0.13
	+10	1.44596	1.82058	20.0866	17.8499	11.1365	+0.02
	−10	1.45534	1.82971	20.2035	17.9679	11.1321	−0.02
	−50	1.47257	1.84675	20.426	18.1913	11.1289	−0.05

i. t_s^* and T^* decrease while Q^*, W^* and AT^* increase with increase in the value of the parameter K. In this case, both t_s^* and T^* are highly sensitive and Q^*, W^* and AT^* are moderately sensitive.

ii. t_s^*, W^* and AT^* increase while T^* and Q^* decrease with increase in the value of the parameters δ_0, C_4 and C_5. In these cases, t_s^*, T^*, Q^*, W^* and AT^* are all less sensitive.

iii. t_s^*, T^*, Q^* and W^* decrease while AT^* increases with increase in value of the parameter γ. In this case, both t_s^* and T^* are moderately sensitive and Q^*, W^* and AT^* are less sensitive.

iv. t_s^*, T^*, Q^*, W^* and AT^* all increase with increase in the value of the parameter C_3. In this case, t_s^*, T^*, Q^*, W^* and AT^* are all highly sensitive.

v. t_s^*, Q^*, W^* and AT^* increase while T^* decreases with increase in the value of the parameter D_0. In this case, t_s^*, Q^*, W^* and AT^* are all moderately sensitive.

vi. t_s^*, T^*, Q^* and W^* decrease while AT^* increases with increase in the value of the parameter α_0. In this case, t_s^*, T^*, Q^* and W^* are less sensitive and AT^* is insensitive.

14.6 CONCLUSIONS

Ouyang et al. (2005) formulated the model with the consideration of exponentially declining demand rate, constant deterioration and the concept of shortages. Due to the variation of the process of deterioration, the constant deterioration rate is not widely useful in developing the model. In this paper, the variable deterioration rate along with exponential demand is considered for developing the model. For shortages, the backlogging rate is inversely proportional to the waiting time for the next replenishment. Furthermore, the numerical example and sensitivity analysis of required parameters are taken for optimising the shortage period and the length of the cycle. The proposed model is suitable for extending into three-parameter Weibull distribution and Gamma distribution. Also, we could extend the proposed model by considering the effect of different demand rates and quantity discounts.

REFERENCES

Chang, H. J., & Dye, C. Y. (1999). An EOQ model for deteriorating items with time varying demand and partial backlogging. *Journal of the Operational Research Society*, 50, 1176–1182.

Covert, R. B., & Philip, G. S. (1973). An EOQ model with Weibull distribution deterioration. *AIIE Transactions*, 5, 323–326.

Dave, U., & Patel, L. K. (1981). (T, Si) policy inventory model for deteriorating items with time-proportional demand. *Journal of the Operational Research Society*, 32, 137–142.

Ghare, P. M., & Schrader, G. H. (1963). A model for exponentially decaying inventory systems. *International Journal of Production and Research*, 21, 449–460.

Goyal, S. K., & Giri, B. C. (2001). Recent trends in modeling of deteriorating inventory. *European Journal of Operational Research*, 134, 1–16.

Hollier, R. H., & Mak, K. L. (1983). Inventory replenishment policies for deteriorating items in a declining market. *International Journal of Production Research*, 21, 813–826.

Ouyang, L. Y., Wu, K. S., & Cheng, M. C. (2005). An inventory model for deteriorating items with exponential declining demand and partial backlogging. *Yugoslav Journal of Operations Research*, 15, 277–288.

Raafat, F. (1991). Survey of literature on continuously deteriorating inventory model. *Journal of the operational research Society*, 42, 27–37.

Singh, T., Pattanayak, H., Nayak, A. K., & Sethy, N. N. (2017). An EOQ inventory model with stock dependent demand under permissible delay in payment. *International Journal of Logistics Systems and Management*, 28, 24–41.

Singh, T., Pattanayak, H., Nayak, A. K., & Sethy, N. N. (2021). An optimal policy with three-parameter Weibull distribution deterioration, quadratic demand, and salvage value under partial backlogging. *International Journal of Rough Sets and Data Analysis*, 5(1), 79–98.

Swain, P., Mallick, C., Singh, T., Mishra, P. J., & Pattanayak, H. (2021). Formulation of an optimal ordering policy with generalised time-dependent demand, quadratic holding cost and partial backlogging. *Journal of Information and Optimization Sciences*, 42(5), 1163–1179.

Wee, H. M. (1995). A deterministic lot size inventory model for deteriorating items with shortages and a declining market. *Computers and Operations Research*, 22, 345–356.

Chapter 15

Smart materials advancements, applications and challenges in the shift to Industry 4.0

Aakash Ghosh, Aryan Sharma, and Navriti Gupta

CONTENTS

15.1 INTRODUCTION

The industrial revolution has brought about a great deal of progression within a multitude of technologies. Each industrial revolution can be linked to the progression of technologies. The First Industrial Revolution in 1784 was incepted based on the utility of steam engines and the mechanization of manufacturing processes. The Second Industrial Revolution in 1870 saw the advancement of the manufacturing processes in a systematic, scalable manner through electrical energy-based mass production techniques such as the assembly line, which increased production efficiencies. This was further supported through an efficient division of labor. In 1969, the Third Industrial Revolution resulted in the automation of processes through field-level computing technologies such as Programmable Logic Computer (PLC), which brought about automation within production processes. In 2011, a project undertaken by the German government formed the fundamental concept of Cyber-Physical Systems (CPS) progressing into Cyber-Physical Production Systems (CPPS). This led to the origin of the term "Industry 4.0" based on the previous industrial revolutions. Under the Industry 4.0 concept, production systems can make decisions based on real-time communication [27]. Essentially, CPPS can make "intelligent" decisions. These decisions can then lead to increased efficiency during the production of various specialized components and products. The concept of Industry 4.0 strongly advocates for the integration of manufacturing systems with embedded systems and the ability of production systems to achieve a great deal of automation [21]. Figure 15.1 depicts the timeline of industrial revolutions [5].

DOI: 10.1201/9781003293576-15

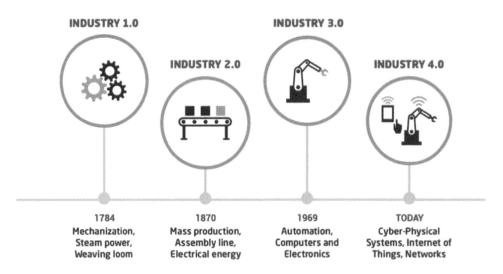

Figure 15.1 Timeline of industrial revolutions 1.0 to 4.0 [5].

Industry 4.0 is a developing area of research in the field of engineering and production. Research has shown that the concept of Industry 4.0 can be intertwined with sustainability, particularly in manufacturing [7]. The United Nations has ascertained sustainability as a key concept that can drive development across the world. This has been further captured in the Sustainable Development Goals and the three dimensions of sustainability, namely economic, social, and environmental. The commonalities between various processes of Industry 4.0 and sustainability exist in the form of life-cycle management, sustainable practices in product design, re-manufacturing, value engineering, and environmental management practices. Emerging concepts that range from digitization of Small Manufacturing Enterprises (SMEs) to closed-loop supply chain solutions have been developed as a result of both Industry 4.0 and sustainability [23]. Figure 15.2 gives a classification of smart materials [8].

Smart materials refer to materials that can change their structure depending on the conditions. They are also termed "shape memory alloys" because of their ability to remember their shape at particular temperature and pressure. Smart materials have garnered the spotlight in recent years as a result of their favorable properties in areas such as aerospace, biomechanics, defense, etc. These materials can be broadly defined as materials which can behave in a systematic and predictable manner when exposed to a stimulus [18]. There is a wide range of materials which come under the smart material umbrella; however, certain alloys have been of key focus due to their ability to re-form upon a temperature-based stimulus. The production and development of smart materials are growing and are expected to become a 73.9 billion dollar industry. The next evolutionary leap for these materials involves the use of a composite blend of two or more materials with a goal to minimize overall costs and mass as well as improve the duration of the active material [20]. The main aim of our research is to coherently present the applications of smart materials and their link with Industry 4.0 while identifying challenges and prospects for the adoption of these materials.

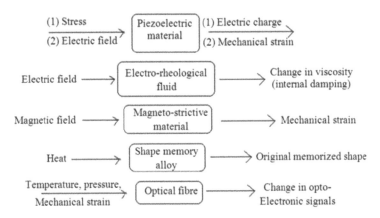

Figure 15.2 Classification of smart materials [8].

15.2 SHAPE MEMORY ALLOYS

Out of the various forms of smart materials, Shape Memory Alloys (SMAs) can be considered as one of the most popular types. The simplicity of their working allows them to be used in multitudes of applications. These materials can be defined as metallic alloys with an ability to recover to their original shape after being plastically deformed. The occurrence of this phenomenon can be devoted to the alloy's phase transformation behavior. The phase change can occur due to a thermal load causing an austenite-to-martensite phase change, and opposingly a load removal causes the alloy to return to its austenite phase [13]. Figure 15.3 depicts the phase transformation diagram of SMAs [13].

SMAs are composed of two or more metals; however, one of the most ubiquitous alloys of this speciality is known as Nitinol. Nitinol is a composition of Nickel and Titanium making it an attractive alloy to several industries. A review study by Sekhar and Uwizeye [19] on the uses of Micro-Electrical Mechanical systems (MEMS) for medical applications discusses the sheer advantages of Nitinol and compliments the alloy's electrical properties, its long fatigue life, and its high corrosion resistance. However, the study also qualifies its praises by highlighting the slow rate at which the wires can be cooled thus limiting the reactiveness of the controller, while also commenting on the greater power requirements in order to enhance the cooling rate. This paper explores the various implementations of SMAs in the industry and focuses primarily on the use of Nitinol due to its commonplaceness. Nitinol is found in the industry in form of wires and can be manufactured to fit different applications.

The main link between Industry 4.0 and smart materials lies in the miniaturization of actuating elements. Ikuta [6] drew comparisons between a wide range of technologies for activations/actuations across all sizes ranging from small Direct Current (DC) motors to large-scale industrial gas turbines. This research found that SMAs have the distinct advantage of low weights coupled with a high power output. This results in a high power to weight ratio while reducing assembly design complexity, making SMAs especially attractive in the field of micro-actuating technology. The greatest advantage

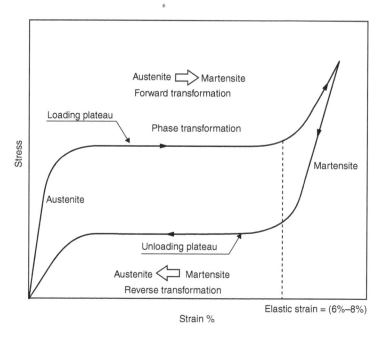

Figure 15.3 Typical phase transformation diagram of SMAs [13].

of miniature actuators lies in the fact that the bandwidths increase and the desirable power density is retained.

15.3   APPLICATIONS OF NITINOL

Nitinol is a prime example of SMAs, and its two prime components are Nickel and Titanium; it finds applications in various industries. The application of these materials is crucial in determining their importance in Industry 4.0. One of the core parts of Industry 4.0 deals with the automation of industry which includes designing as well as manufacturing. The concept of Design For Manufacturing and Assembly (DFMA) will become automated, and any modeling and testing will also be integrated much more effortlessly. Another essential part of Industry 4.0 will be the implementation of additive manufacturing for a range of mass-produced products. By analyzing the major applications, it is possible to realize the recent developments in using SMAs [2].

15.3.1   Aerospace

One of the most popular uses of smart materials lies in the aerospace industry. The significant fields of smart materials which can be seen being used in this industry are SMAs. Due to their simplicity, they can be implemented with relatively greater ease. Costanza and Tata [4] present a review of the possible applications of SMAs in the field of aerospace. SMAs generally have been used as actuators; however, this study delineates more specific roles of SMAs in aerospace. The study explains the use of SMAs in a relatively novel idea of a morphing wing. Morphing wings are wing structures

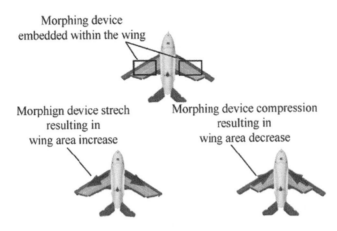

Figure 15.4 Wing area modifications through morphing mechanism [1].

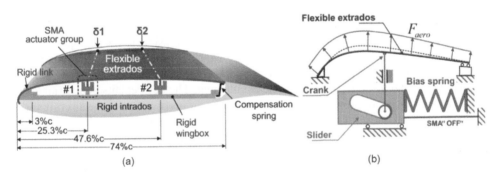

Figure 15.5 (a) Morphing laminar wing concept; (b) morphing mechanism [1].

that can change shape to increase wing area. Two different types of morphing wings are discussed in the study, along with the usage of SMAs in those designs. The first design uses SMA as a morphing device to retract and deploy extensions on the wings, as shown in Figure 15.4 [1].

The second prospective design of the morphing wing includes the alteration of the camber of the wing. The camber is a critical feature of an airplane wing as it has a direct effect on the lift produced at the different cruising velocities.

Figure 15.5 depicts the concept and mechanism used in the morphing wing developed by Brailowski et al. [1]. The wing comprises three main sub-systems: flexible extrados, rigid extrados, and actuators. However, the wing is tested and modeled to operate in subsonic conditions. One main improvement highlighted by Costanza and Tata [4] is the continuous wing surface. Current wing designs have to employ different parts, such as slats and flaps, which have to be mechanically actuated. However, a smoother curve can be obtained using an SMA-based wing structure. A similar study was performed by Laxman and Srivastava [14], who illustrated the use of SMAs in changing the horizontal tail of an aircraft (Figure 15.6).

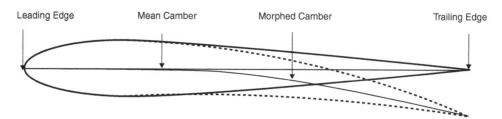

Figure 15.6 Airfoil camber changes [14].

One of the significant improvements of employing SMAs is to reduce the complexity of the flap and slat arrangements used on current wing designs. The study by Laxman and Srivastava [14] successfully performed a numerical analysis on the shape memory effect of a horizontal tail with shape memory alloy strips and noted a deflection of 0.6 mm. Similarly, an extensive study conducted by Simiriotis et al. [21] presented an improved design of SMA-actuated airfoils. The design employed fewer actuators and thus reduced the overall energy costs of the system.

Morphing technology based on SMAs can also be used for various commercial aerospace applications. Calkins and Mabe [3] reviewed the large-scale programs conducted by Boeing in order to validate further and advance the use of SMAs in the aerospace industry. The major results of the Smart Aircraft and Marine Project System Demonstration (SAMPSON) smart inlet program depicted that a set of integrated SMA wires can be used to provide combat aircraft with a variable engine inlet. The key benefits of a variable engine inlet are that it can provide a 20% increase in the radius of action of the aircraft and it could also help enable a subsonic aircraft to carry out supersonic interception operations. The noise from modern-day high bypass ratio turbofan engines can be attributed to mixing ambient air, hot exhaust from the jet, and the stream generated from the fan. Hence, Boeing developed a Variable Geometry Chevron (VGC) to reduce noise. This is a device that uses an SMA-based actuator (60-Nitinol) in order to morph the chevrons as per the desired operation. For take-off, the shape was optimized in order to reduce the noise generated; and for cruising, the shape of the chevrons was optimized for maximum efficiency. This enabled Boeing to achieve a compromise between noise and engine performance.

VGC technology was also used during Boeing's "Variable Area Fan Nozzle" program, under which the noise level of a jet engine can be reduced and its efficiency can be improved. This is achieved by varying the area of the fan nozzle of the jet engine. The nozzle can be operated at a larger diameter at take-off to reduce the jet velocity, which reduces the noise. For cruising, the nozzle diameter can be adjusted as per the operational conditions to reduce fuel consumption.

It is evident that the employment of smart materials is present in the aerospace industry. The use of SMAs can radically change product design and development approaches with much more mechanically simple systems. The commercial aerospace industry has been quite stagnant in design evolution, and complex systems require human input. By employing the mechanisms discussed, the manufacturing process can be automated further. Signs of additive manufacturing can already be noticed in the aerospace industry, where several companies are trying to use 3D printing techniques for manufacturing rockets. 3D printing of standard metals and their alloys is quite difficult, and the same stands true for smart materials. Unoptimized process parameters can lead to defects, which can reduce the shape memory effect as well as lower the fatigue life of SMAs. For such automation to be integrated, it is essential that such manufacturing techniques are thoroughly optimized and the designs are re-evaluated to match the upcoming manufacturing techniques.

15.3.2 Actuators

The advent of SMAs has been incredibly beneficial in developing robotic systems for various applications. The actuators made using SMAs can be used in a wide range of robotic systems. Kheirikhah et al. [10] reviewed and discussed the use of SMA actuators in robotics. They focused on seven different types of robots for their research. After a comprehensive review, the researchers concluded that actuators made using SMAs could replace small-sized robots on account of their greater power to weight ratio and their relative simplicity in construction and assembling. It was also established that the SMA-based actuators are of importance in the field of bio-mimetic, crawler, and flower robots. This is due to the versatile nature of SMAs, due to which they can be engineered to suit movement styles – fast-paced or slow-moving.

Kök et al. [12] presented a review on the applications of smart materials. They discussed the classification of various smart materials and their characteristics. The critical conclusions put forward by the authors suggest that although each type of smart material is essential, it can be inferred that in a broader sense, piezoelectric smart materials and SMAs are especially prominent for actuating applications. The researchers also highlighted the major advantages of smart materials – lesser energy consumption, high sensitivity, great adaptability to changes in the environment, and a small footprint. The reviewers suggest that additives in smart materials represent a potential area in which research can be conducted based on their research.

Two critical takeaways from the aforementioned studies are the ease of assembly and the increased efficiency of utilizing smart materials. Reduced complexity of assembly is crucial for industry 4.0, and this advantage will allow assembly processes to be automated with more ease. Efficiency improvements will allow a reduced effect on the environment, which is crucial for sustainable design development.

15.3.3 Medical applications

In their comprehensive review of SMAs, Mohd et al. [15] noted that over the course of the last 50 years, Nitinol and other SMAs had been used regularly in the medical industry due to their unique properties, particularly excellent resistance to corrosion and their compatibility with the human body. The most crucial property of SMAs, which makes them suitable for use as biomedical implants, is their ability to conform to the shape of the human body.

Morgan [16] established in his review that the need for miniature medical equipment with a high degree of precision, accurate positioning, and reliability means that the use of SMAs will only increase as the demand for such instruments is increasing. These instruments are already being used extensively in neurology, cardiology, orthopedics, and stenting. Figure 15.7 shows the super-elastic trend noticed in SMAs, which

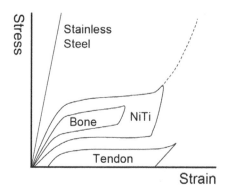

Figure 15.7 Stress-strain behavior of Ni-Ti alloy, human tendon, bone, and stainless steel [16].

largely conforms to the stress-strain behavior noticed in human bones and tendons and is highly useful in stenting applications wherein a narrow tube is inserted inside blood vessels in order to keep the vessels open. The stents made using SMAs are very adaptive to the lumen shape compared to stainless steel stents, which force the blood to flow straight due to their lack of biocompatibility. The stenting instruments industry is expected to be worth around US$ 13 billion by 2025.

A Magnetic Resonance Imaging (MRI) scan uses magnetic fields in short cycles to produce cross-sectional images of the scanning area. The main advantages are that it yields clear images and does not use any form of radiation. Stainless steel has been used in MRI machines; and due to its susceptibility to magnetic fields, the material can often interfere with the resulting image and significantly reduce the clarity of the image. Compared to stainless steel, Nitinol is significantly less susceptible to magnetic fields and, as a result, MRI machines using Nitinol result in more precise images. Since the use of MRI scans is increasing, the use of Nitinol in MRI machines can be a significant development in the field.

Park et al. [17] presented a novel fabric muscle based on SMA springs. Their research can help in the development of artificial limbs. Several medical solutions require the use of non-conforming materials which restrict the patient's movement; however, the use of SMAs can overcome this restriction. Furthermore, intuitive design and implementation of SMAs in the configuration can allow for a larger user base. One of the prime uses where SMA-based muscles can be implemented is the development of exoskeleton and armor reinforcements in the military. Furthermore, the use of such a device could be seen in rescue teams. With the extra strength and capability of the SMA muscle, enhanced armor can be developed in order to remove obstructions and aid in post-havoc situations. For instance, SMAs are utilized in a plethora of humanoid-robot applications, in particular as artificial muscles, as actuators based on SMA wires can closely simulate the characteristics of a human muscle [22]. In another research work by Kikuchi et al. [11], a leg-shaped robot (Leg-Robot) was used by patients with ankle joints. In the design process of the Leg-Robot, the requirements of the age group 75–80 were considered.

The use of SMAs in the medical industry holds a lot of potential because of clear improvements in design. However, the use of SMAs will require improvements in the simulation-based analysis of SMA-based designs. There is a large research potential in predictive modeling of these materials' biomedical parts. Furthermore, with the development of additive manufacturing for smart materials, personalized prosthesis will become more convenient.

15.4 CHALLENGES

The challenges associated with SMAs are widespread, ranging from producing complex shapes to controlling the compositional properties. Some techniques, such as additive manufacturing, a near-net-shape technology that helps control the composition of the alloy, can help in the direct fabrication of complex shapes [25]. The most complex challenge is to retain the property of shape memory alloy after the machinability. Generally, during the machining process, the input parameters such as cutting speed, feed, depth of cut, and machining environment control the output process parameters such as cutting force, cutting temperature, etc. [9]. During the 3D printing process, a strongly exothermic reaction occurs in Ni-Ti powder mixtures during 3D additive manufacturing. It can cause disturbance of the melt pool and cause the nickel in the alloy to evaporate. Selective Laser Melting (SLM) helps in controlling this type of undesired behavior [26].

Industry 4.0 relies on the successful integration of novel manufacturing techniques such as additive manufacturing as it allows production to be more flexible while also reducing scrap. Recent studies have involved machine learning models in predicting the effect of process parameters on the end product. For example, a study by Ye et al. [28] utilized Bayesian Optimization to optimize the process parameters of Laser Beam-Powder Bed Fusion (LB-PBF) for the additive manufacturing of Nitinol. The study models the process with high accuracy but highlights the need to increase convergence iterations and include more process parameters to lower the error further.

15.5 CONCLUSION

The concept of Industry 4.0 demands advanced materials to be used as machine parts. It will reduce the rejection rates and result in higher quality and lower power consumption. Integrating IT technologies with material science will help create a database about different advanced materials and their utilities. Smart materials are the newer generation of materials that are being used in every field of engineering, medical sciences, advanced materials, etc. The review of the applications of SMAs, particularly Nitinol, has highlighted the versatile nature of the materials exhibiting the shape memory effect. As depicted, they can be used for a wide variety of applications ranging from laboratory-based experiments to full-scale commercial operations in the aerospace industry.

Further, these materials have shown a great degree of biocompatibility and conform relatively easily to the shapes of the human body. Hence, they find use across medical and biomechanical applications as well. Overall, the emergence of SMAs can lead to a paradigm shift across industries to adopt these materials to improve accuracy and operational efficiency.

The future belongs to applications of Artificial Intelligence and Machine Learning to the functioning of various devices, which will be fabricated by SMAs and smart materials. Also, 4D printing techniques are being used to fabricate artificial limbs, artificial palms, and artificial robots. Additive manufacturing techniques for SMA parts still display particular challenges [24]. The use of a 3D digital light processing printer enables the production of shape memory polymers parts for flexible electronics. This process works by layer-by-layer photopolymerization of methacrylate semi-crystalline molten macromonomers, which enables rapid fabrication of complex objects and imparts shape memory functionality for electrical circuits [29]. Technologies need to be further developed and researched to integrate SMAs and other smart materials in Industry 4.0.

15.6   FUTURE PROSPECTS

The use of SMAs across a variety of applications has been validated and demonstrated. However, they have not reached their full potential. The key drivers of this are the high cost associated with such materials and the complexity. The following points are some areas that should be further explored in order for SMAs to be implemented into Industry 4.0:

• Computational modeling of SMAs to better predict, understand, and analyze their behavior under testing conditions is essential. Modeling and computation allow new designs and solutions to be tested and optimized. The use of finite element analysis can reduce the number of physical iterations for a particular design. Further development of these methods will aid in the utilization of these materials in Industry 4.0.
• Predictive modeling in manufacturing is another crucial aspect of Industry 4.0 as it helps in foreseeing possible faults during machining or production. The development of more accurate neural network predictive models will allow manufacturing techniques to be used more optimally for SMAs.
• Research and development in the field of actuators with more compact and fast controllers to further enhancing the efficiency of new mechanisms and design. Reduced complexity of design and assembly will allow certain assembly processes to be more automated.
• SMAs have the potential to revolutionize current product design and development principles and open engineers and designers to newer approaches to design problems. A prime field of research will include SMA and additive manufacturing centric design.

From a commercial standpoint, robust communication channels between scientists, engineers, and marketing personnel need to be established to share the knowledge of SMA features across industries. This can also lead to novel applications and faster integration with Industry 4.0.

REFERENCES

1. Brailovski, V., Terriault, P., Georges, T., & Coutu, D. (2010). SMA actuators for morphing wings. *Physics Procedia*, 10, 197–203. doi: 10.1016/j.phpro.2010.11.098.

2. Butt, J. (2020). A conceptual framework to support digital transformation in manufacturing using an integrated business process management approach. *Designs*, 4(3), 17. doi: 10.3390/designs4030017.

3. Calkins, F. T., & Mabe, J. H. (2010). Shape memory alloy based morphing aerostructures. *ASME Journal of Mechanical Design*, 132(11), 111012. doi: 10.1115/1.4001119.

4. Costanza, G., & Tata, M. (2020). Shape memory alloys for aerospace, recent developments, and new applications: a short review. *Materials*, 13(8), 1856. doi: 10.3390/ma13081856.

5. Glistau, E., & Coello Machado, N. I. (2018). Industry 4.0, logistics 4.0 and materials-Chances and solutions. In *Materials Science Forum* (Vol. 919, pp. 307–314). Trans Tech Publications Ltd. doi: 10.4028/www.scientific.net/MSF.919.307.

6. Ikuta, K. (1990). Micro/miniature shape memory alloy actuator. *Proceedings, IEEE International Conference on Robotics and Automation*. doi: 10.1109/robot.1990.126323.

7. Jamwal, A., Agrawal, R., Sharma, M., & Giallanza, A. (2021). Industry 4.0 technologies for manufacturing sustainability: a systematic review and future research directions. *Applied Sciences*, 11(12), 5725. doi: 10.3390/app11125725.

8. Kamila, S. (2013). Introduction, classification and applications of smart materials: an overview. *American Journal of Applied Sciences*, 10(8), 876–880. doi: 10.3844/ajassp.2013.876.880.

9. Kaya, E., & Kaya, İ. (2019). A review on machining of NiTi shape memory alloys: the process and post process perspective. *The International Journal of Advanced Manufacturing Technology*, 100(5), 2045–2087. doi: 10.1007/s00170-018-2818-8.

10. Kheirikhah, M. M., Rabiee, S., Edalat, M. E. (2011). A review of shape memory alloy actuators in robotics. In: Ruiz-del-Solar, J., Chown, E., Plöger, P.G. (eds) RoboCup 2010: Robot Soccer World Cup XIV. RoboCup 2010. Lecture Notes in Computer Science, Vol. 6556. Springer, Berlin, Heidelberg. doi: 10.1007/978-3-642-20217-9_18.

11. Kikuchi, T., Oda, K., Yamaguchi, S., Furusho, J. (2010). Leg-robot with MR clutch to realize virtual spastic movements. *Journal of Intelligent Material Systems and Structures*, 21, 1523–1529. doi: 10.1088/1742-6596/149/1/012060.

12. kök, M., Qader, İ., Dağdelen, F., & Aydoğdu, Y. (2019). A review of smart materials: researches and applications. *El-Cezerî Journal of Science and Engineering*, 6(3), 755–788. doi: 10.31202/ecjse.562177.

13. Lago, A., Trabucco, D., & Wood, A. (2019). An introduction to dynamic modification devices. *Damping Technologies for Tall Buildings*, 107–234. doi: 10.1016/b978-0-12-815963-7.00004-x.

14. Laxman Hattalli, V., & Srivatsa, S. (2018). Wing morphing to improve control performance of an aircraft - an overview and a case study. *Materials Today: Proceedings*, 5(10), 21442–21451. doi: 10.1016/j.matpr.2018.06.553.

15. Mohd, J. J., Leary, M., Subic, A., & Gibson, M. (2014). A review of shape memory alloy research, applications and opportunities. *Materials & Design*, 56, 1078–1113. doi: 10.1016/j.matdes.2013.11.084.

16. Morgan, N. (2004). Medical shape memory alloy applications—the market and its products. *Materials Science and Engineering: A*, 378(1–2), 16–23. doi: 10.1016/j.msea.2003.10.326.

17. Park, S., Kim, U., & Park, C. (2022). A novel fabric muscle based on shape memory alloy springs. Soft robotics. Retrieved 22 January 2022, doi:10.1089/soro.2018.0107.

18. Rogers, C. A. (1988). Smart materials, structures, and mathematical issues; *U. S. Army Research Office Workshop*, Virginia Polytechnic Institute and State University, Blacksburg, Sept. 15, 16, 1988, Selected Papers. United States.

19. Sekhar, P., & Uwizeye, V. (2012). Review of sensor and actuator mechanisms for bi-oMEMS. *MEMS for Biomedical Applications*, 46–77. doi: 10.1533/9780857096272.1.46.

20. Shehata, N., Abdelkareem, M., Sayed, E., Egirani, D., & Opukumo, A. (2021). Smart materials: the next generation. *Reference Module in Materials Science and Materials Engineering*. doi: 10.1016/b978-0-12-815732-9.00062-0.

21. Simiriotis, N., Fragiadakis, M., Rouchon, J., & Braza, M. (2021). Shape control and design of aeronautical configurations using shape memory alloy actuators. *Computers & Structures*, 244, 106434. doi: 10.1016/j.compstruc.2020.106434.

22. Sohn, J. W., Kim, G. W., & Choi, S. B. (2018). A state-of-the-art review on robots and medical devices using smart fluids and shape memory alloys. *Applied Sciences*, 8(10), 1928. doi: 10.3390/app8101928.

23. Stock, T., & Seliger, G. (2016). Opportunities of sustainable manufacturing in Industry 4.0. *Procedia CIRP*, 40, 536–541. doi: 10.1016/j.procir.2016.01.129.

24. Van Humbeeck, J. (2018). Additive manufacturing of shape memory alloys. *Shape Memory and Superelasticity*, 4(2), 309–312.

25. Walker, J., Andani, M. T., Haberland, C., & Elahinia, M. (2014, November). Additive manufacturing of Nitinol shape memory alloys to overcome challenges in conventional Nitinol fabrication. In *ASME International Mechanical Engineering Congress and Exposition* (Vol. 46438, p. V02AT02A037). American Society of Mechanical Engineers. doi: 10.1007/s40830-018-0174-z.

26. Wang, C., Tan, X. P., Du, Z., Chandra, S., Sun, Z., Lim, C. W. J., Tor, S. B., Lim, C. S., & Wong, C. H. (2019). Additive manufacturing of NiTi shape memory alloys using pre-mixed powders. *Journal of Materials Processing Technology*, 271, 152–161. doi:10.1016/j.jmatprotec.2019.03.025.

27. Xu, X., Lu, Y., Vogel-Heuser, B., & Wang, L. (2021). Industry 4.0 and Industry 5.0—Inception, conception and perception. *Journal of Manufacturing Systems*, 61, 530–535. doi: 10.1016/j.jmsy.2021.10.006.

28. Ye, J., Yasin, M., Muhammad, M., Liu, J., Vinel, A., & Slvia, D. et al. (2021). Bayesian process optimization for additively manufactured nitinol. Retrieved 20 January 2022, doi: 10.26153/tsw/17555.

29. Zarek, M., Layani, M., Cooperstein, I., Sachyani, E., Cohn, D., & Magdassi, S. (2016). 3D printing of shape memory polymers for flexible electronic devices. *Advanced Materials*, 28(22), 4449–4454. doi: 10.1002/adma.201503132.

Chapter 16

Virtual Try On

A study on the changing dimensions of jewellery retailing through augmented reality

Hareesh Kumar U. R. and Ambeesh Mon S.

CONTENTS

16.1 INTRODUCTION

An immersive experience in digitally centred marketing strategies using AR, Virtual Try On, etc., drives the consumer to try innovations in modern-day marketing. AR can be stated as a set of technologies where the actual world can be seen as an augmented world by incorporating computer graphics and sensory inputs (Jamwal et al., 2021). As a game-changing strategy, it helps the business firms to offer a new wave of experience in Industry 4.0. Virtual Try On allows an individual to visualize how a product or service will look like in reality. In the shopping process, a consumer spends much time selecting a product because they are more concerned about the suitability of the product, and the question of "does it fit" influences them to make rational choices. As e-commerce grows and develops, the jewellery industry has also

DOI: 10.1201/9781003293576-16

turned into online retailing. However, only 4–5% of the jewellery sales happen online (Banuba, 2019). This is because a physical store provides the opportunity to wear a piece of jewellery in real, and hence the purchase decision can be made based on that. But in online commerce, it is not possible. AR offers a solution for this inconvenience to the customers. Jovarnik et al. (2021) observed that Virtual Try On applications can persuade the consumer more about the appeal and craft of the luxury goods in an interactive way better than a shopping cart website. AR in the jewellery market helps customers to select the right piece from different designs. It can help jewellery retailers in designing products, marketing, and advertising (Zhang, 2021). But it is important to note that the application of AR is not only limited to online buyers but also helps offline customers who are visiting a physical retail store for purchase. Jewellery belongs to speciality or luxury goods (Maxmillian, 2017). Customers need a special purchasing effort when making a speciality good or luxury good purchase. The customer also needs a reasonable time for this special purchasing effort. The Artificial Intelligence-based Virtual Try On method aims to help the customer to simplify this purchasing effort and reduce product selection time (Hwangbo et al., 2020). The Virtual Try On method is becoming widespread in the jewellery industry today, allowing the consumer to wear their favourite jewellery before purchasing, with the help of artificial intelligence, whether sitting at home or going directly to the retail store. However, Kakkar and Chitrao (2020) argue that demographic constructs create barriers for consumers to try innovations in jewellery purchasing. In the context of the COVID-19 pandemic, sanitation and hygiene became essential practices. In this case, it was difficult for the consumer to select the jewellery only after directly wearing it. Jewellery retailers were not also able to encourage more customers to wear the same jewellery for trial due to hygiene concerns. In this context, the Virtual Try On method of jewellery ornaments becomes relevant. Hence, the objective of this research is to assess the potential of Virtual Try On applications in the modern-day jewellery business. The study also examines how top branded jewellery retailers in India have incorporated Virtual Try On in their marketing strategies. This article is narrated based on secondary data.

16.2   STATEMENT OF THE PROBLEM

The jewellery industry in India is the one that represents India's unique cultural diversity and traditional beliefs. At the same time, gold and jewellery products play a crucial role in the economic conditions of the people of the country. The growth of technology is creating revolutionary changes in all industries. Technology is also making pertinent changes in the gems and jewellery industry, ranging from jewellery manufacturing to after-sales service. The most significant changes have happened with the advent of artificial intelligence. With the advent of artificial intelligence, innovative ways of combining AR and virtual reality have made unprecedented changes in the jewellery industry, and the new concept of customized marketing came into play. This article examines the changes brought about by artificial intelligence in the field of jewellery marketing and how strongly they are reflected in the Indian jewellery market. This article focuses on how the concept of Virtual Try On has transformed jewellery retailing into Industry 4.0 standards.

16.3   OBJECTIVES OF THE STUDY

1. To evaluate the potential of the Virtual Try On method in jewellery retailing
2. To examine the Virtual Try On centred marketing strategies of Indian branded jewellery retailers

16.4   RESEARCH METHODOLOGY

The study has been designed as a fact-finding study, and a descriptive approach is followed. The study examined the various existing literatures on the role of AR in the jewellery industry in the global as well as Indian scenarios. The present study is conducted based on secondary data collected from research journals, websites, annual reports, and newspapers. In order to include the updated data in the research, the researcher has considered those articles which have been published after 2017. In the case of journal articles, only those articles that have been published in international peer-reviewed journals were referred for the present research. A total of ten articles published in different peer-reviewed journals were used. In addition, data from five different blogs were also used for the present research.

16.5   OPERATIONAL DEFINITIONS OF KEY TERMS

16.5.1   Virtual Try On

It is the way in which a jewellery consumer tries on a product like a ring or a necklace or a bangle through mobile or other digital devices equipped with a camera and AR-enabled technique.

16.5.2   Augmented reality

It is a technology that enables to superimpose a system-generated image on a consumer's view of the real world. Hence, it provides a composite view to the consumer in the 3D dimension.

16.5.3   Personalized marketing

In the context of the present article, personalized marketing refers to the marketing strategy by which jewellery retailers deliver individualized content to the consumers through data analytics based on users' interaction with the internet and with the use of modern automation technology.

16.6   DISCUSSION AND ANALYSIS

The application of AR in the jewellery industry brings consumers a multidimensional view of jewellery products with the help of Virtual Try On applications. These applications now replace expensive paper catalogues; hence, the consumer can feel a different product experience. While making jewellery purchasing, the consumers can be deeply familiar with the jewellery design and can easily decide whether it suits them without any complex and dubious thoughts. AR-enabled mobile applications offer Virtual Try On so that the consumers can virtually examine and customize their desired product in a 3D dimension. Virtual Try On applications combine the real and virtual world using sensors and offer interactive real-time experience in the 3D dimension. The performance, usability, and value addition to the business make a Virtual Try On application the best (Bellalouna, 2021). Through "Virtual Try On," retailers are offering a new way to enhance the consumer experience that probably makes better sales and increases user satisfaction (Borromeo, 2020). Virtual Try On applications provide a "Try before buy" option for customers without visiting the retail shop. Virtual Try On can be either through smartphone applications or other devices like AR-enabled intelligent mirrors. This method offers a personalized and realistic product experience to the customers. Hence, the customer can try out different varieties of products instantly and examine related product suggestions and alternatives, consequently encouraging customer experience and sales. For online customers, this technology helps them to try out a product from a remote location and then add it to their preferred shopping cart (mirrAR, 2021). AR-enabled applications offer consumers an opportunity to customize the desired product as per their specific demands. Hence, these applications help the customer to change the size, shape, design, etc., as per their requirement. These devices are also helpful in developing personalized marketing strategies. Virtual Try On applications have numerous track points for generating customer-related data at every point of search by collecting a detailed record of consumer search history and the kind of jewellery products the customer searched for. AR-oriented applications or smart mirrors work with the camera device embedded in them. The customer needs to pick a favourite product from the digitally displayed catalogue, like a ring, chain, bracelet, or earring. Then the camera attached to the consumer's device launches, and the app will scan the face, hand, or neck of the consumer. Next, on the screen of the consumer's device, a 3D model of the selected product is superimposed on the face or hand or neck of the consumer (Lodon, 2021). Jewell, CaratLane, MirrAR, JewelAR, Trillion, etc., are some of the most popular Virtual Try On applications available in the Google Play Store.

AR-based applications also offer customers a unique service on product packaging. French jeweller "La mome bijou" offers a virtual packaging system for their customers. In this system, the customer can scan a particular Quick Response (QR) code attached along with the product package to provide additional information about the products or the jeweller. However, even without a QR code, just superimposing the product package on the AR app, customers can access more content on the product like graphics, videos, texts, etc. (Morozova, 2020).

16.6.1 Virtual Try On centred marketing strategies of Indian branded jewellery retailers

In 2017, a fashion tech start-up named Styledotme launched India's first artificial intelligence-enabled MirrAR Business to Business Software As A Service (SAAS). MirrAR is an exclusive Virtual Try On application for Indian branded jewellery sellers. Later, the famous Indian branded jewellery retailers like Tanishq, Kalyan Jewellers, PC Jeweller, and Amrapali Jewels integrated their business with MirrAR to offer Virtual Try On service to the customers. Among Indian retailers, Tanishq has brought a revolutionary change in marketing through Virtual Try On technology. Tanishq launched Virtual Try On kiosks at Bengaluru and Delhi airports. Later, Kalyan Jewellers's online retail platform, "Candere," started a customer Virtual Try On service. PC Jeweller has begun to use AR-enabled digital catalogues in their 11 showrooms in New Delhi. CaratLane is a popular virtual jewellery try-on application in India. It has a partnership with Tanishq for providing Virtual Try On service to the customers. Reliance Jewels has also introduced Virtual Try On systems for their customers (Kumbhat and Mishra, 2020).

One of the imperative factors that can challenge the Virtual Try On applications is the lack of quality in presentation. Even though the product is good and looks elegant, if the presentation through the Virtual Try On application is poor, the customer selection rate will fall (Kumar and Mashooq, 2021). So, only those applications that use advanced image intelligence techniques, 3D modelling, and advanced computer vision technology are competitive. Perfect positioning and accurate colour reflection of ornaments are significant for excellent Virtual Try On applications. Data privacy is also a concern in Virtual Try On technology. Some customers may not be ready to disclose private information. They may think that such Virtual Try On applications capture personal data from their smartphones or computers. Hence, it is necessary to ensure the consumers' data privacy (Plotkina and Saurel, 2019).

16.7   FUTURE RESEARCH DIRECTIONS

This part deals with the future research directions regarding the use of Virtual Try On applications in business. From the analysis of existing studies, the researcher has found the following areas to be addressed in future studies.

16.7.1   User's attitude and satisfaction with Virtual Try On applications of luxury brands

The various existing studies cover the potential marketing opportunities of Virtual Try On applications. It would give more insights if future research is carried out on the user's attitude and satisfaction regarding the use of Virtual Try On applications. Such studies will help to add more insights into consumer innovativeness.

16.7.2   Data privacy and Virtual Try On applications

One of the significant challenges that emerged from the modern-day digital marketing approach is data privacy. Virtual Try On applications are vulnerable to data privacy since the companies use the consumer data generated from applications for their personalized marketing strategies. Hence, there is a need to include studies regarding the Virtual Try On applications and their effect on data privacy.

16.7.3   Virtual Try On and its implications in smart retailing

Smart retailing is the integrated retailing of smart technologies with traditional methods. In the era of Industry 4.0, smart retailing is gaining momentum quickly. Future research can be carried out on the significance of Virtual Try On applications or AR in smart retailing.

16.8   CONCLUSION

Artificial intelligence plays a vital role in business, medical, engineering, manufacturing, etc. With the growth of artificial intelligence, the market became more digitalized, and innovative technologies such as AR and virtual reality began to be widely used in business. COVID-19 pandemic further strengthened the possibilities of AR. During the period of COVID-19, jewellers switched to a new technology called Virtual Try On, an AR-centred marketing strategy. This technology is beneficial for the customers to wear the jewellery virtually from a remote place of their choice without going to the physical store and customizing it according to their requirements. Although not very popular in India, branded jewellery retailers encourage Virtual Try On customer service through mobile applications, Virtual Try On kiosks, etc. The jewellery industry in India is likely to exploit the potential of AR more aggressively in the upcoming days. AR is also a game-changing strategy in product packaging. Various studies have found that the use of AR Virtual Try On applications can enhance the consumer experience. Virtual Try On applications can represent the appeal and craft of jewellery products in a more interactive way. But at the same time, it is imperative to prevent the possibility of breach of consumer privacy through such applications. Future research can be carried out on users' attitudes and satisfaction with Virtual Try On applications, their effect on data privacy, their implications in smart retailing, etc.

REFERENCES

Banuba. (2019). 7 virtual jewelry try ons to wear jewelry in 3D. Medium. https://banuba.medium.com/7-virtual-jewelry-try-ons-to-wear-jewelry-in-3d-7b546cfbdc93.
Bellalouna, F. (2021). Digitization of industrial engineering processes using the augmented reality technology: Industrial case studies. *Procedia CIRP*, 100, 554–559. doi: 10.1016/j.procir.2021.05.120.
Borromeo, E., Delos Santos, A. S., & Tomas, S. C. (2021). AR virtual try-on application of MAC Cosmetics and purchase intention mediated by AR experience among Gen Y beauty consumers. Retrieved November 3, 2021 from https://animorepository.dlsu.edu.ph/etdb_dsi/7.
Hwangbo, H., Kim, E. H., Lee, S. -H., & Jang, Y. J. (2020). Effects of 3D virtual "Try-On" on online sales and customers' purchasing experiences. *IEEE Access*, 8, 189479–189489. doi: 10.1109/ACCESS.2020.3023040.
Jamwal, A., Agrawal, R., Sharma, M., & Giallanza, A. (2021). Industry 4.0 technologies for manufacturing sustainability: A systematic review and future research directions. *Applied Sciences*, 11(12), 5725. doi: 10.3390/app11125725.
Javornik, A., Duffy, K., Rokka, J., Scholz, J., Nobbs, K., Motala, A., & Goldenberg, A. (2021). Strategic approaches to augmented reality deployment by luxury brands. *Journal of Business Research*, 136, 284–292. doi: 10.1016/j.jbusres.2021.07.04.

Kakkar, S., & Chitrao, D. P. V. (2020b). Consumer resistance to innovations in ornamental gold jewellery. *Academy of Marketing Studies Journal*, 25(1). https://www.abacademies.org/articles/consumer-resistance-to-innovations-in-ornamental-gold-jewellery-9701.html.

Kumar, A., & Mashooq, M. (2021). The impact of Covid-19 on weddings: An Indian context. *International Journal of Research in Business Studies*, 6(1), 111–124.

Kumbhat, A., & Mishra, J. K. (2020). Advancements in technologies in retail-an Indian perspective. *Parishodh Journal*, 9(2), 1574–1582.

Lodon, L. (2021). Virtual try-on is more than a pandemic trend and these brands are reaping the rewards. Retrieved November 3, 2021, from https://www.forbes.com/sites/lelalondon/2021/05/20/virtual-try-on-is-more-than-a-pandemic-trendand-these-brands-are-reaping-the-rewards/?sh=4d31cf656c82.

Maximilian, C. (2017). Classifications of products - Convenience, shopping or specialty product? Marketing-Insider, Retrieved November 3, 2021, from https://marketing-insider.eu/classifications-of-products/marketing-insider.eu/classifications-of-products/.

mirrAR. (2021). Virtual Try-On & why it's the next big thing in e-commerce. Retrieved November 3 2021, from https://www.mirrar.com/blogs/what-is-virtual-try-on#:~:text=It%20enables%20customers%20to%20try.

Morozova, A. (2019). Augmented reality for packaging. Medium. Retrieved November 3 from https://medium.com/@info_35021/augmented-reality-for-packaging-abb07001adfb.

Plotkina, D., & Saurel, H., (2019). Me or just like me? The role of Virtual Try-on and physical appearance in apparel M-retailing. *Journal of Retailing and Consumer Services*, 51, 362–377.

Zhang, Y. (2021). Virtual Try-on: The next big thing in luxury business. Retrieved November 3, 2021, from https://hapticmedia.com/blog/virtual-try-on/.

Chapter 17

Analysis of the barriers of blockchain adoption in Land Record System

Priyanshu Sharma, Ramji Nagariya, Mithilesh Kumar, and Bharat Singh Patel

CONTENTS

17.1 INTRODUCTION

In all ages,the new-normal is set by a few disruptive technologieshaving long-lasting implications and benefits. Blockchain is one such exponential technology, for which continued growing need is felt wide scale. Although blockchain technology initially emerged in the finance sector to carry out a financial transaction in a trusted manner, the governments and companies are exploring the benefits of this technology for improved governance and ensuring transparency in various other areas also [1].

In any country, land ownership or registry system contains land records and it constitutes an important department of state governance [2]. In most countries, land records and registries date back to the colonial era, under the influence of fraudsters and fake certificates, plagued with miscoordination between departments, delayed update of records, and ambiguity in the documents' predicaments [3]. These drawbacks in the existing, traditional land record system often give rise to corruption and legal disputes, requiring a large number of government resources from judiciary and law enforcement agencies to settle the issues [4]. Blockchain is capable to sort out the issues related to the land registry system. Transforming to a blockchain-based land registry would prima facie involve: digitization of all records – old or new; rectifying anomalies

DOI: 10.1201/9781003293576-17

in existing records; developing cryptographic proofs for property transactions; designing protocols for data authentication and authorizing changes to records; lawful and regulatory frameworks for data safety; and implementing a phased transition from the blockchain free to the new blockchain-based system [5]. Scaling up the paradigms and prototypes to address the issues in, and absence of, valid authorizations with data will be the way forward in reforming land records management.

Countries are looking ahead to establish themselves as a leader in blockchain technology [6]. In India, more than 10 years ago, Digital India Land Records Modernization Programme (DILRMP) was initiated by Govt. of India to digitize land records. The programme was targeted to bring the country's land records to the level seen in developed nations,where information and its usage arekept in easy-to-access central, virtual record rooms with regular updates in real time [4]. Several countriestook the initiative to promote and encourage the use of blockchain technology and research in the area of blockchain technology. Despite the promotion of DILRMP, the digitalization of the land record system is not prepared for the adoption of blockchain technology in it [7]. There are certain factors which resist the adoption of blockchain in the land record system. The present study is aimed to find the answers of the following reseatch questions (RQs):

RQ1: What are the barriers to blockchain technology adoption in the land record system?

RQ2: How these challenges/barriers are interrelated with blockchain technology in the land record system?

The format of the chapter is structured as follows: Section 17.2 communicates the review of literature; Section 17.3 presents the research framework; Section 17.4 presents the case study; Section17.5 presents the results and discussion of this study; and Section 17.6 discusses the conclusion.

17.2   REVIEW OF LITERATURE

A disruptive technology like blockchain transacts and exchanges digital information and assets of goods and financial resources in distributed networks with a high degree of transparency as well as security [8,9]. The blockchain is a self-regulating, still auditable, trust-free, and tamper-proof system [10]. This technology is an implementation in the processes of asset registry, information exchange, and regulation of tangible and intangible assets [9] and can be used by both governments and different stakeholders. Apart from the financial transactions in the banking sector, where it was initially used,the various areas where public administration can use blockchain technology areland and property registration and records, digital identity and authentication, agri-management, public health records, etc.[11].

In developing nations, records of land and property are difficult to access and maintain and under constant threat of natural or artificial man-made disasters as they are kept in a centralized location. To increase the dependability, trustworthiness, and transparency of the land record registration system, the governments of developing countries are exploring new possibilities to digitize land records [12]. Honduras, Malta, UAE, Russia, Sweden, Singapore, China, Spain, Switzerland, Denmark, the

United Kingdom, Ukraine, Australia, and the United States of America have taken initiatives related to blockchain [13–15].

17.2.1   Barriers of blockchain usage for land records

In land registry and record system, there are inter-organizational (as it involves many government and public departments), systems-related, intra-organizational, and external (public acceptance and adaptability) barriers. The adoption of blockchain technology has the potential to solve the issues related to these barriers easily. In this section the barriers to blockchain adoption are mentioned.

a. **Disputes:** To start with, each and every aspect of land records should be indisputable. Though blockchain can ensure integrity and indisputability, it cannot resolve existing differences. For example, it cannot provide proper information about the land whose area is not measured correctly or whose ownership is controversial or whose value is missing [16].

b. **Simultaneous for all the phases:** Due to the existence of poor quality land records, the introduction of blockchain to all land parcels is unsuitable. First of all, it should be introduced on lands having the least issues. This transition can be applied to government agencies, both central and state. Government Land Information System (GLIS) has been created by the central government to demonstrate details of all land holdings [17].

   Private owned lands can also be can records,Banks and financial institutes commence due attentiveness to validate the information in land records and rectifies the existing differences.Blockchain enhances the security of land records by reducing the risk of lending against property. National e-governance Services Ltd (NeSL), a service utility, can be used in blockchain for collateral land. Subsidized interest rates can be provided initially to incentivize the borrowers. Blockchain secure land records commandsuperiority in terms of lower interest rates and higher liquidity which could allure land owners to seek the security [18].

   This powerful technology can be easily used by the selected participants to seek transition-related information by using proper incentives.

c. **Data digitization and correctness:** The land registry system involves a massive amount of registration documents to be stored on central databases that facilitate the transaction for the trading of land title. This system is prone to various types of manipulation and alterations, due to corrupt employees in the registry office. Sorting out these issues involves many overheads in the form of time, storage, and cost. The land registry departments have taken initiatives to take advantage of the information and communications technology (ICT) for increasing openness and transparency by distributing the data from a single database to a distributed database. However,the correctness of records is still a question,and it is still prone to alterations and scams. The blockchain also offers ways to track the data of previous transactions [19].

d. **Data centralization, lack of privacy, and safety threats:** Currently, blockchain architectures rely on centralized,but third-party assisted transaction models, where most networks are recognized, legalized, and attached via central servers.

Considering the scale-up at the state or national level, the central cloud servers become an issue and thus may be a great failure point that can distort the total network. Further, safe communication frameworks either just exchange all user data without the owner's permission without considering user privacy or reveal ambiguous data, not understandable by the applicant. A growing number of self-driving functions may often lead to serious accidents because of the security breach by malicious software [20].

e. **Public acceptance, fear, and digital readiness:** In every state, various stakeholders at different levels are attached to the land record and registry system. There is a unique perception of various stakeholders regarding the positives and negatives of blockchain [21]. Despite the much-hailed advantages of blockchain, many participants are afraid of dependency on blockchain operators [22].

f. **Infrastructure and high-speed internet:** Although India is in the stages of transforming to digitize working, still more than half of its population is rural-dweller. Apart from cost considerations, blockchains are complex and a block chain-based land transaction system requires additional resources. For creating mutual commitment and shared vision, cooperation and adjustment are required in the business strategies of various stakeholders which may create a challenge for smooth adoption of blockchain [23].

17.2.2 Proposed framework for the land registry system

In the process of land trading, the buyer and seller have to sign a pre-agreed smart contract and further send the request for transfer to the registry offices. The registry offices verify the individuality of the seller as well as of the buyer and then check the record of the land title with the reviewer and available records. As soon as the process of 'validation and verification' is complete, the blockchain system enables the 'financial transaction' and collects the different processing charges. This helps in transferring the land ownership, and a certificate is provided to the buyers. At the same time, the records are updated at various levels. The verification of seller and buyer is done with the help of pre-agreement ID that stores the seller and buyer ID and agreement details.

17.2.2.1 New transaction block

Once the transaction process is complete and ownership records are created, these records need to be updated at all departments to verify the records. The transactions are recorded in separate ledger accounts by the parties and historical transactions are recorded in Distributed Ledger Technology (DLT) of that property. All the transactions will be checked and verified by the concerned offices which are already stored in the distributed ledger. The newly created transaction blocks are verified by the validating node and then it is broadcasted to all the nodes within the DLT network. The other validating nodes will verify and agree on the generated node with the help of a consensus mechanism and then finally the new block is added to the blockchain.

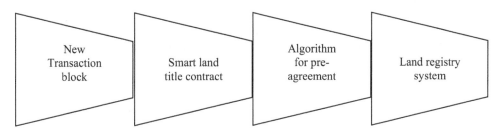

Figure 17.1 Framework for blockchain adoption in the land registry system.

17.2.2.2   Smart land title contract

The pre-agreement contract process between buyer and seller has been shown below that requires the following details given in Figure 17.1. The internal contents have been discussed in detail as follows:

a. **Identity:** It stores the information (ID,name,sign) of all the users in the system and uses the know your customer feature for the verification of legal identity.
b. **Title:** It stores and updates the information related to the Title on the blockchain like ID, address, status, and the current owner.
c. **Deed:** It stores and manages the information related to Deed like ID, seller ID, buyer ID, and payment status.
d. **Agreement:** It storesvarious legal agreements,an inspection report by the surveyor, and purchase agreements; and it is digitally signed for deed transitions.
e. **Electronic signature:** It storesdigital signatures of all the participants in the system and allows storing multiple signatories acting in different capacities in the same documents.

17.2.2.3   Algorithm for pre-agreement

The pre-agreement contract is between buyer and seller based on IDs and payment details on the amount of land. The basic requirement of generating title is based on property/owner ID and payment record. Further, taxation ID will be generated based on the property ID and rate of taxation as per revenue department rules. The payment ID will be null until verified by the registry office and payment is done by the buyer. Once the registry office verifies all the details of the land, ownership records, dues, etc., the buyer will initiate the payment. After the verification of payment details, the registry office will create the payment approval and start the transfer process.

17.3   RESEARCH METHODOLOGY

Interpretive Structural Modelling (ISM) has been used to make an inter-relationship among the barriers/enablers [24],which are poorly communicated and unstructured [25]. The decision regarding relatedness between a complex set of factors is done using the ISM method. The steps of ISM [24] are as follows:

1. Making a Structural Self-interaction Matrix (SSIM) is the first step in showing the direction of the contextual relationship among the identified enablers.
2. With the help of V, A, X, and O, the initial reachability matrix is constructed. The transitivity concept is used to design the final reachability matrix.
3. The variables are classified in various levels after reachability matrix.
4. Transitivity is required to be removed and a diagraph is constructed.
5. Diagraph is used to construct the ISM model which represents the relative importance of variables.

17.4   CASE STUDY

This study focuses on the barriers faced by the employees and workers of development authorities of Rajasthan, Haryana, Uttar Pradesh, and Madhya Pradesh. Only the officers, employees, and workers were contacted for the collection of responses. The reason for collecting responses from senior personnel was the awareness level of senior authority on the system and the barriers faced by the development authorities of different states and workers. Barriers related to the adoption of blockchain in the land record system were discussed. Four officers,seven employees, and nine workers from different development authoritiesagreed to participate in this study. For this kind of study, a sample size of 20 is recommended by Arnold et al. (2011) [26], which defines the suitability of the sample size used in this study. Due to the confidentiality clause, the names of the participants of the organization and the profiles of the experts are not disclosed. The responses were collected from August 2021 to October 2021. The identified barriers were also presented to them; and after their consent, the barriers were finalized. The aim and purpose of the study were discussed with them. The experts provided their responses to barriers in the ISM sheets. Their responses were aggregated and the final responses are mentioned in the SSIM. Further, these responses were analysed by using the ISM method and the results and findings are discussed in the next sections.

The meaning of the notations and the purpose of the study was explained and were requested to construct the contextual relationship amongst the barriers. The experts made an SSIM with the help of pair-wise relationship. The meanings of the notations used are as follows:

V=barriers i help in achieving barriers j; A = barriers j help in achieving barriers i; X=internal barriers i and internal barriers to each other; O = internal barriers i and j are independent.

Table 17.1 SSIM for barriers of blockchain adoption in the land revenue system

Barriers	B9	B8	B7	B6	B5	B4	B3	B2	B1
Disputability (B1)	O	O	O	O	V	X	O	O	I
Data digitalization (B2)	O	O	O	O	X	O	X	I	
Data corrections (B3)	O	V	O	A	O	O	I		
Centralization (B4)	O	O	O	X	O	I			
Safety threats (B5)	O	O	V	O	I				
Public acceptance (B6)	O	O	O	I					
Digital readiness (B7)	O	O	I						
Infrastructure (B8)	O	I							
Internet speed (B9)	I								

Note(s): Here, SSIM=Structural Self-interaction Matrix; Table 17.1 presents the SSIM for the internal barriers. The SSIM is obtained from the inputs of the experts.

Table 17.2 Initial reachability matrix for barriers of blockchain adoption in the land revenue system

Barriers	B9	B8	B7	B6	B5	B4	B3	B2	B1
Disputability (B1)	0	0	0	0	I	I	0	0	I
Data digitalization (B2)	0	0	0	0	I	0	I	I	0
Data corrections (B3)	0	I	0	0	0	0	I	I	0
Centralization (B4)	0	0	0	I	0	I	0	0	I
Safety threats(B5)	0	0	I	0	I	0	0	I	0
Public acceptance (B6)	0	0	0	I	0	0	I	0	0
Digital readiness (B7)	0	0	I	0	0	0	0	0	0
Infrastructure (B8)	0	I	0	0	0	0	0	0	0
Internet speed (B9)	I	0	0	0	0	0	0	0	0

Table 17.3 Final reachability matrix for barriers of blockchain adoption in the land revenue system

Barriers	B9	B8	B7	B6	B5	B4	B3	B2	B1	Driving power
Disputability (B1)	0	0	1a	1a	I	I	0	0	I	5
Data digitalization (B2)	0	1a	1a	0	I	0	I	I	0	5
Data corrections (B3)	0	I	0	0	1a	0	I	I	0	4
Centralization (B4)	0	0	0	I	1a	I	1a	0	I	5
Safety threats (B5)	1a	0	I	0	I	I	1a	I	0	5
Public acceptance (B6)	0	1a	0	I	1a	0	I	1a	0	5
Digital readiness (B7)	I	0	I	0	0	0	0	0	0	2
Infrastructure (B8)	0	I	0	0	0	0	0	0	0	2
Internet speed (B9)	I	0	I	0	0	0	0	0	0	2
Dependence power	3	4	5	3	6	2	5	4	2	

Table 17.4 Iteration I

Barriers	RS	AS	IS	Level
Disputability (B1)	1, 4, 5, 6, 7	1, 4	1,4	
Data digitalization (B2)	2, 3, 5, 7, 8	2, 3, 5, 6	2,3,5	
Data corrections (B3)	2, 3, 5, 8	2, 3, 4, 5, 6	2,3,5	
Centralization (B4)	1, 3, 4, 5, 6	1, 4	1,4	
Safety threats (B5)	2, 3, 5, 7, 9	1, 2, 3, 4, 5, 6	2,3,5	
Public acceptance (B6)	2, 3, 5, 6, 8	1, 4, 6	6	
Digital readiness (B7)	7, 9	1, 2, 5, 7, 9	7,9	I
Infrastructure (B8)	8	2,3,6,8	8	I
Internet speed (B9)	7,9	5,7,9	7,9	I

Note: In the first iteration, B7, B8 and B9 are assigned level I. In the second iteration, B2, B3 and B5 are assigned level II.

Table 17.5 Iteration II

Barriers	RS	AS	IS	Level
Disputability (B1)	1, 4, 5, 6	1, 4	1,4	III
Data digitalization (B2)	2, 3, 5	2, 3, 5, 6	2,3,5	II
Data corrections (B3)	2, 3, 5	2, 3, 4, 5, 6	2,3,5	II
Centralization (B4)	1, 3, 4, 5, 6	1, 4	1,4	III
Safety threats (B5)	2, 3, 5	1, 2, 3, 4, 5, 6	2,3,5	II
Public acceptance (B6)	2, 3, 5, 6	1, 4, 6	6	III

Note: In the third iteration, B1, B4 and B6 are assigned level III.

For constructing the final reachability matrix, the concept of transitivity is used. Transitivity tries to establish the relationship amongst the variables which are not directly related but are related indirectly. For understanding the concept of transitivity, let us suppose that there are three variables V1, V2, and V3. And V1 and V2 are directly related. Similarly, V1 and V3 are directly related. Then, the concept of transitivity says that all variables are interrelated. Table 17.3 depicts the transitivity in Ref. [25].

The hierarchical levels are decided by classifying the barriers. The classification of barriers at different levels is carried out with the help of the Intersection Set (IS) of the Reachability Set (RS) and the Antecedent Set (AS). The RS contains thebarriers itself and the additional barriers being helped by the barriers. The same barriers in the IS and the RS indicate that the level number should be assigned to the barrier and this levelled barrier should be removed from further levelling of the barriers. Until all barriers are levelled, the same method should continue. Table 17.4 shows the first iteration and first level assigned to barriers B7, B8, and B9.

ISM model is constructed from the achieved levels of the barriers (Figure 17.2).

Out of nine barriers, public acceptance (B6), centralization (B4), and disputability (B1) were identified as the most influencing barriers having high driving power. Data correction (B3), data digitalization (B2), and safety threats (B5) were having moderate driving power and low dependence power and, therefore, were identified in the middle

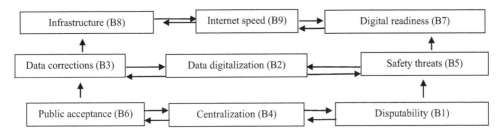

Figure 17.2 ISM model for the barriers of blockchain adoption in the land registry system.

of the ISM diagram. Infrastructure (B8), internet speed (B9), and digital readiness (B7) were identified as barriers having low driving power and low dependency on other barriers.

17.5 DISCUSSION

The study identified key barriers that affected the performance and the land record system of differentdevelopment authorities. The major barriers were identified from a rigorous literature review and were finalized after the consent from the officers, workers, and employees. A total of nine barriers were acknowledged. The participants helped in categorizing the barriers into higher driving power and lower driving power. Several senior-level experts agreed to participate in this study. Their responses were collected and analysed by using the ISM methodology. ISM created the hierarchical structure of the barriers.

The study focuses on the barriers that arose and affected the performance related to the land record system of four state development authorities. During the research, the authors found that employees faced barriers to using the land record system. As no data corrections could be provided to the organizations, many officers/employees found it difficult to connect with data digitalization. Especially, the new officers faced this problem significantly. Due to the barriers of internet speed and infrastructure, the officers of the organization could not connect with the public. Centralization of data and public acceptance related to the land record system also come under a significant barrier. Blockchain does not support records of disputable lands and also poses safety threats. Due to a lack of public awareness,digital readiness is also termed as an important barrier. Therefore, it can be concluded that these barriers have high driving power and low dependence power. Such barriers affect the barriers of the linkage region also and the corresponding changes in the barriers of the linkage region should be noticed carefully. The study will enable the decision-makers (officers) of different development authorities in understanding the barriers related to the adoption of the land record system.

The study also makes significant theoretical and managerial contributions and these are mentioned in the subsequent subsections.

17.5.1 Theoretical implications

The study identifies the barriers that affect the land record system. The barriers are classified into higher driving power and lower driving power. These barriers could be considered to be a significant theoretical contribution. The identification of the barriers and the contextual relationship among these barriers can be considered to be the major theoretical contribution. The study also identified the critical barriers majorly affecting the land record system.

17.5.2 Practical implications

There are numerous practical implications of the study. This research will help the decision-makers of different development authorities in identifying and understanding the real barriers faced by the organization. The understanding of the barriers will enable them to handle these barriers effectively and, therefore, can implement a better quality land record system. The hierarchy of the barriers will enable the decision-makers in identifying the critical barriers that majorly affect the land recording process and ultimately affect the reputation of the organization.

17.6 CONCLUSION

In this chapter, a total of nine barriers are identified by taking into consideration a systematic literature review and opinions of officers. The ISM method helps to draw a better understanding of the hierarchical association of the identified key barriers. 'Public acceptance' (B6), 'Centralization' (B4), and 'Disputability' (B1) were identified as the most influencing barriers having high driving power.'Infrastructure' (B8), 'Internet speed' (B9), and 'Digital readiness' (B7) were identified as barriers having low driving power and low dependency on other barriers.

This study has its own limitations and, in future, these can be explored by the researchers. As the study is dependent on a limited sample, the findings of the study cannot be generalized. In future, the researchers can take a large sample size and use Structural Equation Modelling (SEM) to generalize the findings of the research. Also, the same type of study can be conducted in different countries and the results could be compared with the findings of this study. This will help in identifying new barriers and a comparative study will help in identifying the country- or region-specific barriers in the blockchain adoption in the land registry system. The barriers depend on the technology and people's attitude; therefore, the barriers can be considered dynamic in nature. So, future studies may consider these new barriers and can adopt other methods to explore these barriers.

REFERENCES

1. Nakamoto, S. (2008). Bitcoin: A peer-to-peer electronic cash system. Retrieved April 9, 2017, from Bitcoin - Open source P2P money. https://bitcoin.org/bitcoin.pdf.

2. Bhatnagar, S., & Chawla, R. (2007). Computerizing land records for farmer access. Ending Poverty in South Asia, 219.

3. Sinha, S. (2018). How Andhra Pradesh is emerging as India's block chain hub. https://analyticsindiamag.com/how-andhra-pradesh-is-emerging-as-indias-blockchain-hub/.

4. Thakur, V., Doja, M. N., Dwivedi, Y. K., Ahmad, T., & Khadanga, G. (2019). Land records on blockchain for implementation of land titling in India. *International Journal of Information Management*, 52, 101940. https://doi.org/10.1016/j.ijinfomgt.2019.04.013.

5. Chandra, V., & Rangaraju, B. (2017). Block chain for property, A roll out roadmap for India, India Institute. http://indiai.org/blockchain-handbook/.

6. Krupa, K. S., & Akhil, M. S. (2019). Reshaping the real estate industry using blockchain. In Sridhar, Y., Padma, M. C., & Radhakrishna Rao, K. A. (eds.), *Emerging Research in Electronics, Computer Science and Technology* (pp. 255–263). Springer, Singapore.

7. Mclean, S., Relton, D., & Thatcher, B. (2019). India announces a new national framework for blockchain. A Blog by Becker McKenzie. 6–12–2019 Last Assessed: 21–08–21.

8. Abeyratne, S.A., & Monfared, R.P. (2016). Blockchain ready manufacturing supply chain using distributed ledger. *International Journal of Research in Engineering and Technology*, 5(9), 1–10.

9. Ølnes, S., Ubacht, J., & Janssen, M. (2017). Blockchain in government: Benefits and implications of distributed ledger technology for information sharing. *Government Information Quarterly*, 34(3), 355–364.https://doi.org/10.1016/j.giq.2017.09.007.

10. Atzori, M. (2015). Blockchain technology and decentralized governance: Is the state still necessary? Available at SSRN 2709713. http://ssrn.com/abstract=2731132.

11. Behara, K.G. (2019). Why block chain technology is the need of the hour for land management. https://www.expresscomputer.in/blockchain/why-blockchain-technology-is-the-need-of-the–for-land-management/43907/.

12. Benbunan-Fich, R., & Castellanos, A. (2018). Digitization of land records: From paper to blockchain.

13. Jun, M. (2018). Blockchain government-a next form of infrastructure for the twenty-first century. *Journal of Open Innovation: Technology, Market, and Complexity*, 4(1), 7. Available: http://link.springer.com/article/10.1186/s40852-018-0086-3.

14. Ojo, A., & Adebayo, S. (2017). Blockchain as a next generation government information infrastructure: A review of initiatives in D5 countries. In Ojo, A., & Millard, J. (eds.), *Government 3.0–Next Generation Government Technology Infrastructure and Services* (pp. 283–298). Springer, Cham.

15. Shen, C., & Pena-Mora, F. (2018). Block chain for cities—a systematic literature review. *IEEE Access*, 6, 76787–76819.

16. Hughes, L., Dwivedi, Y. K., Misra, S. K., Rana, N. P., Raghavan, V., & Akella, V. (2019). Block chain research, practice and policy: Applications, benefits, limitations, emerging research themes and research agenda. *International Journal of Information Management*, 49, 114–129.

17. Kim, S. K., & Huh, J. H. (2020). Block chain of carbon trading for unsustainable development goals. *Sustainability*, 12(10), 4021.

18. Asharaf, S., & Adarsh, S. (Eds.). (2017). *Decentralized Computing Using Block ChainTechnologies and Smart Contracts: Emerging Research and Opportunities: Emerging Research and Opportunities*. IGI Global, Pennsylvania.

19. Viriyasitavat, W., Da Xu, L., Bi, Z., & Sapsomboon, A. (2020). Block chain-based business process management (BPM) framework for service composition in industry 4.0. *Journal of Intelligent Manufacturing*, 31(7), 1737–1748. https://doi.org/10.1007/s10845-018-1422-y.

20. Dorri, A., Steger, M., Kanhere, S. S., & Jurdak, R. (2017). Block chain: A distributed solution to automotive security and privacy. *IEEE Communications Magazine*, 55(12), 119–125. https://doi.org/10.1109/MCOM.2017.1700879.

21. Deshpande, A., Stewart, K., Lepetit, L., & Gunashekar, S. (2017). Distributed ledger technologies/blockchain: Challenges, opportunities and the prospects for standards. *Overview Report the British Standards Institution (BSI)*, 40, 40.
22. Kim, S. (2018). Blockchain for a trust network among intelligent vehicles. In *Advances in Computers*, Vol. 111, pp. 43–68. Elsevier. https://doi.org/10.1016/bs.adcom.2018.03.010.
23. Reyna, A., Martín, C., Chen, J., Soler, E., & Díaz, M. (2018). On blockchain and its integration with IoT. Challenges and opportunities. *Future Generation Computer Systems*, 88, 173–190. https://doi.org/10.1016/j.future.2018.05.046.
24. Warfield, J. N. (1974). Developing interconnection matrices in structural modeling. *IEEE Transactions on Systems, Man, and Cybernetics*, SMC-4(1), 81–87. https://doi.org/10.1109/TSMC.1974.5408524.
25. Nagariya, R., Kumar, D., &Kumar, I. (2021). Enablers to implement sustainable practices in the service only supply chain: A case of an Indian hospital. *Business Process Management Journal*, ahead-of-print. https://doi.org/10.1108/BPMJ-10-2020-046925.
26. Arnold, J.M., Javorcik, B.S. and Mattoo, A. (2011). Does services liberalization benefit manufacturing firms? Evidence from the Czech Republic, *Journal of International Economics*, 85, 136–146. doi:10.1016/j.jinteco.2011.05.002.

Chapter 18

The concept of Industry 4.0

Role of ergonomics and Human Factors

S. Ranjan, V. Roy, and Navriti Gupta

CONTENTS

18.1 INTRODUCTION

In today's world, technology is evolving at such a rapid pace that robots will completely replace humans in their own competition. Industry 4.0 is the HRC, and it comprises mainly nine technologies as Artificial intelligence, Machine learning Robotic Process Automation (RPA) or HRC, Edge computing, Quantum computing, Virtual reality and augmented reality, Block-chain, Internet of things, 5G, and Cyber security which completely changes the future of humans [1]. Also, advancement in the industry leads to complexity due to technological development and these are trying to set some new kinds of demand for the practices and processes used in companies management as well as for the personal competency and skill-set [2].

The Fourth Industrial Revolution or Industry 4.0 conceptualizes the advancement in industrial technologies with the help of artificial intelligence, gene editing, and advanced robotics in modern industries. This industrial revolution has made the industries independent of human beings and also it has increased the efficiency and production in industries to many folds. In essence, the Fourth Industrial Revolution is the trend toward automation and data exchange in manufacturing technologies

DOI: 10.1201/9781003293576-18

and processes which include cyber-physical systems (CPS), Iota, industrial internet of things, cloud computing, cognitive computing, and artificial intelligence. Some biggest trends are Smart factory, Predictive maintenance, 3D printing, and Smart sensors [3].

Moreover, the attention is on ergonomics. Ergonomic principles tend to provide some scientific correlation which is related to the understanding of the interaction between the human beings and other elements of a system (machines) that makes use of different methodology, ideas, data, and ethics to develop a strategy that can grow the overall performance of human well-being in Industry 4.0 [4].

Human resources are the most important asset for any kind of organization. For a very long time, there is a constant need and effort to improve the safety and current working conditions of individuals in any type of organization or industry. The major aim of safe working conditions is to provide a comfortable working situation to enable workers to carry out their laborious and hefty tasks/jobs with ease and fewer efforts.

The advancement in ergonomics combined with the latest innovations in the field of science and technology has contributed a lot toward the massive industrialization globally. Continuous inspection, selection, and creation of tools, machines, and work processes are being developed in the modern working environment. For instance, over time hammers, axes and plows have been further developed bringing about the more efficient and effective usage of these tools in performing their various tasks and operations. Recent research in ergonomics is tilted specifically toward the logistics in innovation activities as these form one of the key obtainable areas in the field of ergonomics [5]. It is crucial to acknowledge the ergonomic assessment flawlessly, broadly, and for the most part rapidly.

18.2   ADVANCEMENT IN TECHNOLOGIES IN INDUSTRIES WITH REGARD TO TECHNOLOGY ADVANCEMENT COMPONENTS

18.2.1   Ergonomics

With the help of continuous research and technological advancement, industries have become fully automated and the human resources requirement has reduced a lot. The complex nature of these technologies poses an inability for humans to operate efficiently along with the danger to their lives. This arises the need for advancements in technologies used in Industry 4.0 that is favorable to human beings. Advancement in technologies in industries about ergonomics is a concept that emphasizes the development of technologies that are comfortable, easy to operate, and most importantly safe to be operated by a human being.

The use of principles of ergonomics in Industry 4.0 helps to provide a comfortable working environment for workers in an industry, improving the individual safety of workers in the industry. When the workers work in a safe and comfortable working environment, then it makes their lives safer. From Table 18.1, the aforementioned technological methods help to reduce certain health and risk hazards when employed as compared to when an individual would have carried out manufacturing processes manually [4]. This will result in strengthening the manufacturing processes in Industry 4.0, which results in even more efficient, reliable, and smooth functioning of the industries.

Table 18.1 Application of technology in Industry 4.0

Technology	Usage
Computer-Aided Design (CAD)	It helps the user to design the products before starting the production in either 2D or 3D. It enables the development, modification, and optimization of the design process.
Integrated engineering system	It involves the use of IT support systems in the development of various products and manufacturing processes in the manufacturing industry.
Automation with sensors	Sensors are devices used for automatic monitoring of process with little or no human intervention.
Simulations and analysis of virtual models	It helps with the model-based design of systems.

18.3   EVOLUTION OF INDUSTRY 1.0 TO 4.0

As human civilization exists on the earth, the willingness to know will be there; so, we reached an era of Industry 4.0 now. The first revolution of the industry began in the 18th century through the use of steam power and mechanization of production [6].

18.3.1   First Industrial Revolution

The First Industrial Revolution was mainly focused on the transition from hand reduction methods to machines in industries. The revolution started with new chemical manufacturing and iron production processes, followed by the development of machine tools and then the rise of mechanized factory systems [7].

In terms of employment, the value of output, and capital invested, the textile industries were the dominant industry of the industrial revolution. Modern production methods were also first used by the textile industries [6]. The first country that tasted the industrial revolution was Great Britain. Many technological and architectural innovations originated in British. By coming of the mid-18th century, Britain became the world's leading commercial nation [8].

During the First Industrial Revolution, the important technological developments were textile manufacturing, iron industries, steam power, machine tools, chemicals, cement, gas lightning, glass making, paper machine, agriculture, mining, transportation, and other developments [7]. The social effects of the industrial revolution were that the factory systems were implemented all over the continents, and suddenly the standards of living become better for people. Some other social effects were literacy and industrialization, clothing and consumer goods, population increase, urbanization, and effects on women and family life. There were effects on labor conditions and also on environmental conditions and nationalism [8,9].

18.3.2   Industrial Revolution

The Second Industrial Revolution started around the 1850s in which the steel was used in mass production, from small needle to large manufacturing. In the 1860s, operating

Sir Henry Bessemer invented a new furnace that can convert the molten pig iron into still that too in large quantities; so after the invention, the industrialization grew more rapidly. And in the 1870s, many new inventions took place in which the Bessemer furnace was displaced by the open-hearth furnace [10].

Soon, this industrial revolution grew rapidly to include the various chemical industries and petroleum (refining and distribution) [9]. In the 20th century, the automotive industry was marked by a transition of technological leadership from Britain to the United States and Germany.

The potential of industrialization was further widened due to a reduction in the importance of coal due to the increasing availability of economical petroleum products [11].

Electricity and electrification in electric industries resulted in the beginning of a new revolution. In the 1890s, the rapid industrialization of coal-driven northern Italy was enabled due to the introduction of hydroelectric power generation in the Alps [12].

By the 1890s, the first giant industrial corporations with global interests were formed due to industrialization in these areas as companies like US Steel, General Electric, Standard Oil, and Bayer AG joined the railroad and ship companies on the world's stock markets.

18.3.3   Third Industrial Revolution

In the Third Industrial Revolution, the emergence of another source of energy happened which was nuclear energy. This industrial revolution brought us the rise of telecommunication, computers, electronics, the invention of the internet, and automation of the industrial process; and it also opened another door to exploration in space, research, and biotechnology.

18.3.4   Fourth Industrial Revolution (Industry 4.0)

The Fourth Industrial Revolution or Industry 4.0 conceptualizes the advancement in industrial technologies and methodologies with the help of modern and advanced technologies such as artificial intelligence, gene editing, and advanced robotics in modern industries and processes. This industrial revolution has made the industries independent of human beings and also it has helped in increasing the efficiency and production in industries to many folds by incorporating various modern technologies into industries and production processes.

Industry 4.0 has resulted in the digitalization of products and services by evolving new methods for data collection and analysis of data such as the expansion of an already existing product or the creation of any new product. This helps the companies in modern times to generate data on product use, which eventually helps in product refinement.

This industrial revolution leads to the concept of the "smart factory". The smart factory is the vision that aims toward a production environment in which production facilities and logistics are organized without human intervention. Under this concept, small structured factories are made with a virtual copy of the physical world that doesn't require humans for production. All the manufacturing and production processes are controlled by highly complex machines that use artificial technology and Internet of Things (IoT).

The Fourth Industrial Revolution or Industry 4.0 is mainly tilted toward the automation of industries using modern technologies like artificial intelligence and data exchange in manufacturing industries and processes which include CPS, Iota, industrial internet of things, cloud computing, and cognitive computing. Some biggest trends in Industry 4.0 are Smart factories, Predictive maintenance, 3D printing, Smart sensors, etc [3].

18.3.4.1   Key concepts

Germany was the first country to discover and incorporate the concepts of Industry 4.0 to suit its position as a producer and a technologically advanced country. It was introduced in the year 2011. The current wave of the industrial revolution is called Industry 4.0 or the Fourth Industrial Revolution. Industry 4.0 revolutions are primarily connected to modern technologies and their use in industries; for example, artificial intelligence, digitalization, the internet of things, additive manufacturing, cyber-physical systems, cloud computing, rapid increases in automation in manufacturing industries, and the use of robotics in the manufacturing processes in the industry [2].

In 1800, the mechanization and utilization of mechanical power were used in revolutionizing industries, which started the First Industrial Revolution. Electricity and electrification in industries set the premises for the start of the Second Industrial Revolution and mass production. In the 1960s, with the help of digitalization and the use of microelectronics and automation in manufacturing industries, the Third Industrial Revolution started. The rapid development and advancement of modern technologies like artificial intelligence, information, and communication technologies triggered the Fourth Industrial Revolution [2].

This revolution has led to an increase in the use of information technology, automation, and digitalization in the industry which has made the manufacturing process very efficient because the time consumption has reduced along with an increase in quality as a machine always produces goods of the same quality.

The principles of Industry 4.0 don't consider the HF, which is very crucial for the overall development of a human-friendly working environment.

This revolution has also made manufacturing very complex and exclusive for human beings. The human requirement in industries has decreased due to the revolution. Now, more skilled workers are required for operating such complex manufacturing machines.

The use of robotics has decreased human involvement in manufacturing and production in industries. The human-robot interaction has led to strengthening the manufacturing processes.

18.4   ERGONOMICS AND HUMAN FACTORS (HFs)

Ergonomics is a concept related to comfort, efficiency, and compatibility in the working environment. In every organization or industry, the most valuable asset is human resources. Ergonomics and HFs act as a basic ground that provides the framework for improving the safety, performance, skills, compatibility, and interaction between the revolutionized technology and human beings in manufacturing processes under Industry 4.0. With the introduction of the concept of Industry 4.0, the use of various technologies has led to the development of intelligent manufacturing units which are aided by artificial intelligence, automation, information, and communication technologies. But these advanced industries have neglected the importance to create a human-friendly working environment. Having an HMI in the industry can lead to strengthening the manufacturing processes even more.

For this purpose, we use ergonomics to convert complex technologies into human-friendly ones. This can be done by considering the HF during very preliminary stages like the designing stages.

HFs and Ergonomics is a field of work mainly concerned with the understanding of interactions among humans and other elements of a system as a scientific discipline and it applies theoretical principles, data, and methods, to understand the relationship between human beings and machines. Human factors/Ergonomics (HF/E) has its roots in physical ergonomics, i.e., in anatomical, anthropometric, physiological, and biomechanical characteristics related to physical activities performed by humans [2].

In a technologically advanced industry like Industry 4.0, ergonomics can be applied not only to an individual worker (micro-ergonomics) but also to an organizational level (macro-ergonomics).

In the industries and any corporation, we can reduce all possible occurring dangers significantly by the application of proper ergonomic programs in place. In the industry, during various manufacturing processes, due to the complexity of the latest technologies and sometimes due to lack of skills, various accidents occur which sometimes even cost the lives of human beings. Ergonomics results in lower cases of diseases, illnesses, and deaths [4].

18.5   CASE STUDY: CEIT ERGONOMICS ANALYSIS APPLICATION (CERAA)

CERAA is a mobile application developed by CEIT as a company in collaboration with the Slovak Ergonomics Association and the University of Zilina [13].

With the improvement and advancement in industries with the help of principles of Industry 4.0, there is a need to consider the HF for creating a comfortable and healthy environment for working. Taking one step ahead in the same direction, the mobile application called CERAA was developed. It is a very user-friendly and efficient mobile application developed based on legislation. This mobile application is dedicated to screening the evaluation of spatial conditions and working positions of workers at potentially risky workspaces in industries and various organizations.

In the year 2016, the first version of CERAA was finalized and started getting used by companies such as Lear Corporation Seating Slovakia and many others. This mobile application first went through the phase of professional preparations; after that, it went through practical testing; and then, the pilot deployment into praxis directly to some customers was done. The main users of this mobile application are the manufacturing industries and various industrial firms. However, various observations were also made from the side of health services industries and firms, security technicians, and various technical universities. With the help of CERAA, workers and employees can evaluate the normal working operations and workspaces at which the workers perform their activities/jobs for more than half of the working shifts in the industry [13].

This mobile application can also act as a very efficient tool while preparing a new workspace and also during performing ergonomics principles on a workstation in the industry. One of the greatest advantages of CERAA is that the user of this mobile application need not be an expert in the field of ergonomics. A worker only requires some basic knowledge from the field of ergonomics and human resources, and the detailed workplace designing that the customers acquire during the training period is sufficient for the easy use of this application. For the effective and efficient usage of this mobile application, the customers only need a tablet/mobile, installed application, marker, and should have completed the provided training by the company. In case of any need or emergency, customers could use an extensive user guide provided by the company. This mobile application called CERAA can be operated in three different languages – Slovak, Czech, and English [13].

The major aim of the CERAA application is to evaluate and determine if the current workspace/workstation is risky or not from the view of ergonomics and if a detailed evaluation of the workspace is necessary by the second level tools and design of correctional arrangements, alternatively what health risk threatens workers. This could have finally led to worsening the quality of the production and work efficiency in industries, decreased productivity, and so on. This application is based on the use and application of virtual and augmented reality items [13].

The evaluation of working conditions is the first area of application of the mobile application CERAA for workers on a workstation in the industry. For the development of such a mobile application that can perform ergonomics, information regarding optimum dimensions of the workstation along with average worker body requirements and other information regarding body postures during working action were computed using a series of virtual models of humans and elements of the working environment. From those, various alternates are accurately modeled for different workers regarding working position (sitting, standing), sex, height, and principles of workspace projecting that help in designing or improving the workstation resulting in a comfortable, efficient, and healthy working environment [13].

The advancement of this digital mobile application is a great example of using principles of ergonomics in Industry 4.0. With the increment in automation factories of intelligence and technology under Industry 4.0, the HF got neglected as the technologies became human independent and too complex to be operated by an average skilled worker. This kind of innovation will help in strengthening the human-machine relationship, creating an easy to operate, healthy, and comfortable environment for working. This will increase productivity along with preserving and protecting human resources.

18.6   CONCLUSIONS AND FUTURE SCOPE

The future of Industry 4.0 is very vast, and it should include human ergonomics. The man-machine relationship is a complex relationship being aided by the usage of machine learning and artificial intelligence. However, machines are and always will be dependent on man for their functioning. If the man is dependent fully on machines, then also the ergonomic factors will be neglected.

The concept of Industry 5.0 is also emerging where importance is given to the role of man over machines. And it should not happen that role of humans gets neglected initially and redundant finally.

In the coming future, the main aim of the industrial revolution should be to build technologies that can take the production and efficiency of industries to many folds and without compromising human resources. The focus of future development should be on making human-friendly industries. Industries are the backbone of any country's economy, and they are also the largest jobs providers; so, in the coming future, the industries must focus on those technologies that can improve the production along with creating a friendly man-machine relationship and this can only be achieved by using the concepts of ergonomics in Industry 4.0.

REFERENCES

1. Duggal, N. (2021). Top 9 new technology trends for 2021. Accessed Feb, 22, 2021.
2. Reiman, A., Kaivo-oja, J., Parviainen, E., Takala, E. P., & Lauraeus, T. (2021). Human factors and ergonomics in manufacturing in the industry 4.0 context–A scoping review. *Technology in Society*, 65, 101572.
3. Crafts, N. F., & Mills, T. C. (1994). Trends in real wages in Britain, 1750–1913. *Explorations in Economic History*, 31(2), 176–194.
4. Fayomi, O. S. I., Akande, I. G., Essien, V., Asaolu, A., & Esse, U. C. (2021, April). Advances in concepts of ergonomics with recent industrial revolution. In *IOP Conference Series: Materials Science and Engineering* (Vol. 1107, No. 1, p. 012010). IOP Publ.
5. Yang, L., Hipp, J. A., Lee, J. A., Tabak, R. G., Dodson, E. A., Marx, C. M., & Brownson, R. C. (2017). Work-related correlates of occupational sitting in a diverse sample of employees in Midwest metropolitan cities. *Preventive Medicine Reports*, 6, 197–202.
6. Malthus, T. (1798). *An Essay on the Principle of Population*. London: St. Paul's Church- Yard.
7. Houston, R. A. (1982). The development of literacy: Northern England, 1640–1750. *Economic History Review*, pp. 199–216.
8. Thompson, D. (2012). The economic history of the last 2000 years in 1 little graph'. The Atlantic, 19.
9. Daunton, M. J. (1995). Progress and poverty: An economic and social history of Britain 1700–1850. OUP Catalogue.
10. Ashton, T. S. (1997). The industrial revolution 1760–1830. OUP Catalogue.
11. Tong, J. T. (2016). *Finance and Society in 21st Century China: Chinese Culture versus Western Markets*. Boca Raton, FL: CRC Press.
12. Davies, G. (2010). *History of Money*. University of Wales Press.
13. Gášová, M., Gašo, M., & Štefánik, A. (2017). Advanced industrial tools of ergonomics based on Industry 4.0 concept. *Procedia Engineering*, 192, 219–224.

Chapter 19

Carbon nanotubes as an advanced coating material for cutting tool in sustainable production in Industry 4.0

Navriti Gupta and R. S. Walia

CONTENTS

19.1 INTRODUCTION

The concept of Industry 4.0 advocates for the integration of different industries such as computer, electronics, electrical, chemical, etc., to achieve mass production [1]. The demand of high production manufacturing industries are superior cutting and forming tools, good quality raw materials, and 3D additive manufacturing techniques [2]. Now, high generation cutting tools help in faster cutting, along with a high-quality surface finish. The high quality of machined surface will ultimately result in less rejections of the workpiece, and hence more customer satisfaction. Also, less power is wasted when rejections are less [3,4].

Different types of tools such as carbide tools, cermets, etc., are being used while machining various dies and tools and plastic molds. Researchers are developing newer and newer materials for coating of cutting tools. Coatings such as tungsten carbide, titanium carbide, vanadium, chromium, etc., help in achieving a higher quality of

DOI: 10.1201/9781003293576-19

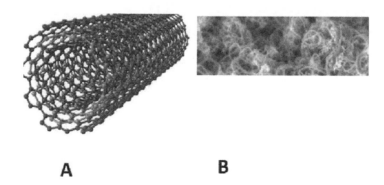

A   **B**

Figure 19.1 (a) Graphene sheet rolled to form CNT, (b) Scanning Electron Microscopy (SEM) image of CNTs [7].

surface finish. CNTs-based nanocomposite coating is a novel area of research. Apart from usual coating materials such as titanium, vanadium, molybdenum, and other advanced coatings, CNT coatings are the latest area of research.

Nanotechnology is an emerging field and a promising field. The other name given to such technology is "Extreme Technology". Here, precision and miniaturization are combined. Nanotechnology covers the molecules having at least one dimension of about 1–100 nm [5]. Rolled graphene sheets form CNTs. They exist in many forms, some of them being single-walled carbon nanotubes (SWCNTs). If they are having many walls one into another, then they are multi-walled carbon nanotubes (MWCNTs). The characteristics of SWCNT are high aspect ratio or length to diameter ratio, enhanced chemical stability, and well-defined atomic structure [6]. However, the synthesis of SWCNTs is a big challenge because of the greater control needed while yielding them. On the contrary, the synthesis of MWCNTs is much easier but their physical and strength properties are inferior to SWCNTs (Figure 19.1).

Various researchers have attempted researches with CNTs-based cutting tools. In experimental research, CNTs were deposited on the high speed steel (HSS) tool. The CNT-coated tool performed in a better way than DLC (Diamond-Like Coating), in terms of tool life and tool failure when machined for the same duration.

In another research work, CNTs-based coating on HSS tool during hard turning of EN 31, resulted in better surface finish and lesser cutting forces [8–10] (Figure 19.2).

In experimental work performed, CNTs were deposited over the HSS cutting tool. There was a significant increase in tool life and lower tool tip temperature [11].

In a research study [12], a dispersion of CNTs, carbon nano-fibers (CNFs), and graphene is prepared in a matrix. The purpose was to obtain a superior carbon-metal mix. So, it has been seen in the literature review that CNTs are emerging as a superior coating substitute. In yet another experimental study, the Ni-Cu metal matrix was reinforced with CNTs to improve corrosion resistance [13]. The advantage is superior enhanced strength, mechanical, thermal, and electrical and strength properties. They combine higher strength along with their light weight.

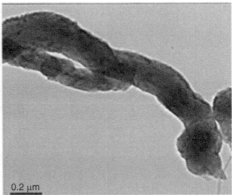

Figure 19.2 SEM and HRTEM (High Resolution Tunnel Electron Microscopy) images of CNTs [8].

19.2 SYNTHESIS OF CNT

The CNTs exist as SWCNTs, DWCNTs (double-walled carbon nanotubes), and MW-CNTs. As far as strength and mechanical properties are concerned, SWCNTs are superior to other CNTs.

Many techniques are being used in the fabrication of CNTs. The techniques involving very high temperatures are PVD, laser ablation, etc., whereas a low temperature technique is Chemical Vapor Deposition (CVD). In CVD technique better temperature control can be obtained, the maximum temperature of process being 800 degree celcius [14]. However, in both of these techniques, we get CNTs with impurities. The type of process selected for fabrication will mainly determine the purity of CNTs obtained. The main type of impure particles in the process comprises mainly carbon-based particles such as amorphous carbon, graphite in nanocrystalline form, fullerenes, etc. Also, the metal particles used as a catalyst during the process, such as Ni, Fe, Co, etc., can alter the purity of CNTs formed. Maintaining purity has its own challenges [15].

19.2.1 Physical Vapor Deposition (PVD) techniques

The PVD involves very high temperatures of the order of 3,000 °C.

19.2.1.1 Pulse laser deposition

Another technique of deposition of CNTs is Pulse Laser. The laser properties such as high temperature, peak power, wavelength of oscillation, energy generation, power consumption, type of workpiece, gas pressure, and distance between laser torch and workpiece determine the quality of CNTs deposited [16].

Figure 19.3 shows laser method of production of CNTs. Here, the graphite target is melted by pulse laser energy and high temperature. This method also generates a very high temperature.

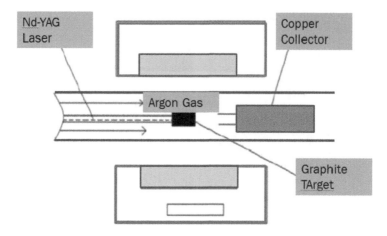

Figure 19.3 Laser method.

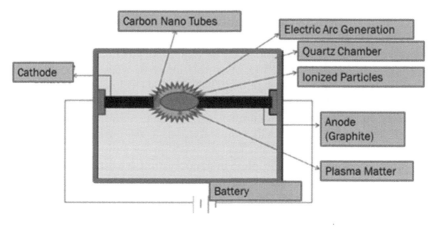

Figure 19.4 Arc discharge method.

19.2.1.2   Arc discharge

In the process of deposition of CNTs by Arc Discharge, very high temperatures (>1,700 °C) are encountered. However, the main advantage is the quality of CNTs obtained. Fewer structural defects are present in the CNTs obtained by this method.

Different catalytic precursors are used for the arc discharge deposition of CNTs [17,18]. For fabrication of SWCNTs, the process requires a transition metal catalyst. However, for the production of MWCNTs, no catalyst is required.

Figure 19.4 shows electric arc deposition method, in which an electric arc is generated between two graphite electrodes. This method involves very high temperatures.

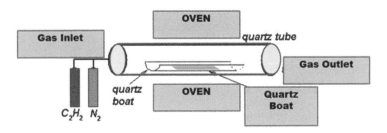

Figure 19.5 CVD method.

19.2.2   Chemical Vapor Deposition (CVD) techniques

A CVD method was invented for nanotube synthesis. In the process, highly aligned CNTs were obtained, which were of 50 nm thickness nanotubes [19].

Here, directional growth is possible up to several micrometers thick as a film. This method is inexpensive also. However, the quality of CNTs obtained is also inferior, mostly MWCNTs.

In Figure 19.5, the CVD apparatus is shown. The input required is a mixture of hydrocarbon gas, acetylene, methane or ethylene gas, which act as a source of carbon. Also, nitrogen is introduced into Quartz Tube, in which the whole reaction will take place. The hydrocarbons were decomposed and nanotubes were formed and deposited on Substrate, at a temperature of 750–890°C [20]. Here, the process is occurring at comparatively low temperatures.

19.3   PURIFICATION OF CNTs

For purification purposes, there are different post-growth treatments, mainly, to eliminate any defects in the tubes produced and their purification also. Separating many tubes that are stuck together or onto the other particles can be done by the ultrasonic bath method [21]. The particle size of impurity makes the purification tough or easy. Smaller particles pose more problems in their elimination. Oxidative treatment, which is possible by a liquid phase treatment in acidic environment, is done for the purification of MWCNTs. Cross-flow filtration, which is a more complicated method, is used for the purification of SWCNTs.

19.4   PROPERTIES OF CNTs

19.4.1   Physical properties

The nanotubes have a very high Young's modulus of elasticity, the average value is 1.8 TPa. The scanning force microscope can be used to measure Young's modulus [22,23].

19.4.2   Electrical properties

The CNTs possess a very good electrical conductivity. When placed in electrical fields, they emit electrons [24].

19.4.3   Thermal properties

CNTs are stable in vacuum up to 3,000 °C. They possess high thermal conductivity. They are stable in the air up to 750 °C. The thermal conductivity is 6,000 W/mK at room temperature which is comparable with nearly pure diamond, which has 3,320 W/mK [25].

19.5   APPLICATIONS OF CNTs

The CNTs have a wide application from engineering to medical to building and construction to defense and space industries.

19.5.1   Genetic engineering

The CNTs display similar properties to nephrons in kidneys in the matter of water transmission and channeling. They can easily be used for highly targeted drug delivery. In gene identification [26], a CNTs-based atomic force microscope can be used.

19.5.2   Aerospace and automotive industry

CNTs are considered for constructing space elevators because of the very high length to diameter (L/D) or aspect ratio. One of the most important characteristics of CNTs is that, in them, scientists have found a combination of high strength and low density. There are multiple ongoing researches on the topic [27,28].

19.5.3   Electronics and chip manufacturing

Field effect Transistors (FET) have been developed using CNTs, which have been termed as Carbon Nanotube FET (CNT-FET). They can operate at low temperatures and resemble silicon devices [29].

19.6   CONCLUSIONS AND FUTURE SCOPE

Taking into account the broader picture of Industry 4.0, mass production has to be achieved considering the concept of sustainability and green technology. Environmental concerns are of important value. The higher and advanced tool coatings will result in faster production rates along with lesser rejections and lesser energy consumption rates. Also, the tool life will be increased, thus avoiding frequent rejections.

The quality will also be enhanced, which will lead to greater customer satisfaction. Also since the rejection is less, wastage is less. This will lead to sustainable manufacturing. CNTs are the emerging materials of the future. Their synthesis is possible by both low-temperature methods (e.g., CVD) and high-temperature methods (e.g., PVD). They are a suitable candidate as tool coating materials. The other tool coatings such as titanium carbide and tungsten carbide are very costly. Since carbon is available in bulk, the coating cost can be low than these in mass production.

Since most small scale machining units are not using high-speed cutting tools, because of their cost, mass production of the CNTs-based coating industry can be a future industry. Also, nanotubes are used in cutting fluids and to decrease the temperatures reached in machining. They also flush away chips and act as a heat sink, thereby protecting the cutting tool edge. The research scope is very high for tool manufacturing from CNTs using powder metallurgy technology.

The CNTs have very good strength and mechanical properties, high Young's modulus of elasticity. They combine strength with light weight. They find their usage in defense, construction, biomechanical, artificial limbs, automotive, and pharmaceutical industries too. CNTs-based drug delivery patches are being manufactured. Their future usage includes developing bullet-proof vests, space elevators, gene identifiers, etc.

REFERENCES

1. Chenrayan, V., Manivannan, C., Velappan, S., Shahapurkar, K., Soudagar, M. E. M., Khan, T. Y., Elfasakhany, A., Kumar, R., Pruncu, C. I. (2021). Experimental assessment on machinability performance of CNT and DLC coated HSS tools for hard turning. *Diamond and Related Materials*, 119, 108568.
2. Shirguppikar, S., Patil, M. (2020). Performance analysis of Multi Wall Carbon Nanotubes (MWCNT) coated tool electrode during machining of titanium alloy (Ti6Al4V). In *International Manufacturing Science and Engineering Conference* (Vol. 84263, p. V002T06A017). American Society of Mechanical Engineers.
3. Suzuki, T., & Konno, T. (2014). Improvement in tool life of electroplated diamond tools by Ni-based carbon nanotube composite coatings. *Precision Engineering*, 38(3), 659–665.
4. Jamwal, A., Agrawal, R., Sharma, M., Kumar, A., Luthra, S., Pongsakornrungsilp, S. (2021). Two decades of research trends and transformations in manufacturing sustainability: a systematic literature review and future research agenda. *Production Engineering*, 1–25.
5. Mamalis, A. G., Vogtländer, L. O. G., Markopoulos, A. (2004). Nanotechnology and nanostructured materials: trends in carbon nanotubes. *Precision Engineering*, 28, 16–30.
6. Gommans, H., Alldredge, J., Tashiro, H., Park, J., Magnuson, J., Rinzler, A.G. (2000). Fibers of aligned single-walled carbon nanotubes: Polarized Raman spectroscopy. *Journal of Applied Physics*, 88(5), 2509–2514.
7. http://www.barc.gov.in/technologies/cnt/index.html.
8. Gupta, N., Agrawal, A. K., & Walia, R. S. (2019). Soft modeling approach in predicting surface roughness, temperature, cutting forces in hard turning process using artificial neural network: an empirical study. In *International Conference on Information, Communication and Computing Technology* (pp. 206–215). Springer, Singapore.
9. Gupta, N., Walia, R. S., & Agrawal, A. K. (2020). Robust Taguchi based optimization of surface finish during hard turning EN 31 with carbon nanotubes-based nano-coated tip. In R. Agrawal et al. (eds.), *Recent Advances in Mechanical Engineering* (pp. 289–298). Springer, Singapore.

10. Gupta, N., Agrawal, A. K., & Walia, R. S. (2020). Taguchi based analysis of cutting force in hard turning EN 31 with indigenous developed Carbon Nano Tubes coated insert. *Materials Today: Proceedings, 25*, 827–832.

11. Chandru, M., Selladurai, V., & Venkatesh, C. (2021). Multiobjective performance investigation of CNT coated HSS tool under the response surface methodology platform. *Archives of Metallurgy and Materials, 66*.

12. Kumar, A. (2018). Methods and materials for smart manufacturing: additive manufacturing, internet of things, flexible sensors and soft robotics. *Manufacturing Letters, 15*, 122–125.

13. Heragh, M. F., Eskandarinezhad, S., Dehghan, A. (2020). Ni-Cu matrix composite reinforced with CNTs: preparation, characterization, wear and corrosion behavior, inhibitory effects. *Journal of Composites and Compounds, 2*(4), 123–128.

14. He, Z. B., Maurice, J. L., Lee, C. S. Cojocaru, C. S., Pribat, D. (2010). Nickel catalyst faceting in plasma-enhanced direct current chemical vapor deposition of carbon nanofibers. *Arabian Journal of Science and Engineering Section B, 35*(1C), 11–19.

15. Kruusenberg, I., Alexeyeva, N., Tammeveski, K., Kozlova, J., Matisen, L., Sammelselg, V., Solla-Gullón, J., Feliu, J. M. (2011). *Carbon, 49*, 4031–4039.

16. Arora, N., Sharma, N. N. (2014). Arc discharge synthesis of carbon nanotubes: comprehensive review. *Diamond & Related Materials, 50*, 135–150.

17. Parkansky, N., Boxman, R. L., Alterkop, B., Zontag, I., Lereah, Y., Barkay, Z. (2004). *Journal of Physics D: Applied Physics, 37*, 2715–2719.

18. Tsai, Y. Y., Su, J. S., Su, C. Y., He, W. H. (2009). Production of carbon nanotubes by single-pulse discharge in air. *Journal of Material Processing Technology, 209*(9), 4413–4416.

19. Li, W. Z., Xie, S. S., Qian, L. X., Chang, B. H., Zou, B. S., Zhou, W. Y., et al. (1996). Large-scale synthesis of aligned carbon nanotubes. *Science, 274*, 1701–1703.

20. Xie, S., Li, W., Pan, Z., Chang, B., Sun, L. (2000). Carbon nanotube arrays. *Materials Science Engineering A, 286*(1), 11–15.

21. Ebbesen, T. W., Ajayan, P. M. (1992). Large scale synthesis of carbon nanotubes. *Nature, 358*, 220–222.

22. Hoa, L. T. M. (2018). Characterization of multi-walled carbon nanotubes functionalized by a mixture of HNO_3/H_2SO4. *Diamond and Related Materials, 89*, 43–51.

23. Wong, E. W., Sheehan, P. E., Lieber, C. M. (1997). Nanobeam mechanics: elasticity, strength, and toughness of nanorods and nanotubes. *Science, 277*, 1971–1974.

24. Rinzler, A. G., Hafner, J. H., Nicolaev, P., Lou, L., Kim, S. G., Tomanek, D., et al. (1995). Unraveling nanotubes: field emission from an atomic wire. *Science, 269*, 1550–1553.

25. Collins, P. G., Avouris, P. (2000). Nanotubes for electronics. *Scientific American*, 38–45.

26. Sa, V. (2011). *Highly conductive carbon nanotube fibers for biomedical applications* (Doctoral dissertation, Clemson University).

27. Smitherman, Jr, D. (2000, September). Space elevators-Building a permanent bridge for space exploration and economic development. In *Space 2000 Conference and Exposition* (p. 5294).

28. Edwards, B. C. (2000). Design and deployment of a space elevator. *Acta Astronautica, 47*(10), 735–744.

29. Dmitriev, A., Lin, N., Weckesser, J., Barth, J. V., & Kern, K. (2002). Supramolecular assemblies of trimesic acid on a Cu (100) surface. *The Journal of Physical Chemistry B, 106*(27), 6907–6912.

Chapter 20

Integrating AI with Green Manufacturing for process industry

Shatrudhan Pandey and Abhishek Kumar Singh

CONTENTS

20.1 INTRODUCTION

Integration of the Internet of things (IoT) with AI and ML, has become more popular in the process industry in recent years due to the Industry 4.0 revolution, as shown in Figure 20.1. The Gross Domestic Product (GDP) of India is increasing at an annual rate of around 7% over the last decade, and India is emerging as one of the biggest economies. Leading government and private entities throughout the world are recognizing India as the next global manufacturing hub. In terms of worldwide chemical sales, the Indian chemical and petrochemical industries are placed sixth in the world and first in Asia, with more than 80,000 chemical items manufactured in India accounting for 2.5% of global sales. The petrochemical industries can have a significant impact on the development of the country's economy, but it has negative aspects like chemical hazards, health, and environmental consequences. In the era of digital media, accidents in the process industry are getting more attention. GM is one of the viable solutions to make the environment clean and green. The role of an intelligent system is useful for the integration of information, preventive risk assessment, decision-making, and early warning for better safety and risk management implementation. Chemical industries in India influence every aspect of the country's economic development and so play a vital role in influencing people's lives and overall chemical production. Indian chemical industries are classified as large, medium, and small enterprises which manufacture significant petrochemicals, alkali chemicals, organic and inorganic chemicals, insecticides, dyes, and pigments.

DOI: 10.1201/9781003293576-20

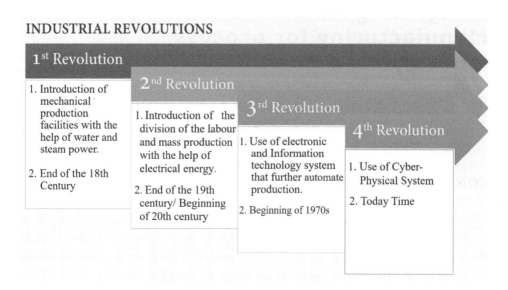

Figure 20.1 Industrial revolution.

The chemical industry is large and accounts for 8.8% of India's manufacturing Gross Value Added (GVA) and 1.4% of the country's total GVA. This is crucial in the manufacturing sector since chemicals and chemical products are used in a wide range of industries, including food and drinks, textiles, leather, metal extraction and processing, petroleum refining, medicines, rubber, and so on. According to National Industrial Classification for Financial Year (FY) - 2019, Ministry of Statistics and Programme Implementation, Government of India, coke, petroleum, rubber chemical, and related products value addition from petrochemical products is 32% in the FY 2019, as shown in Figure 20.2.

As a result, the manufacturing sector's Index of Industrial Production (IIP) is linked to the country's chemical manufacturing. In the previous 5 years, the IIP of the total manufacturing and chemicals manufacturing category has increased. In terms of value, specifically, chemical industries have risen at an outstanding pace of roughly 11.7% in the previous 5 years [1].

20.2 CHALLENGES OF PROCESS SAFETY IN THE CONTEXT OF GREEN MANUFACTURING (GM)

Accidents in the early 1980s drew attention to process industries into the mainstream due to poor safety. In the 1990s, risk management system was developed to prevent safety-related issues.

The chemical sector is one of the safest, yet the public perception of its safety has deteriorated. Perhaps, this is because chemical plant accidents can be dramatic and get a lot of media attention. The public frequently links the chemical sector with environmental and safety issues, giving it a bad image. Hazards should be eliminated

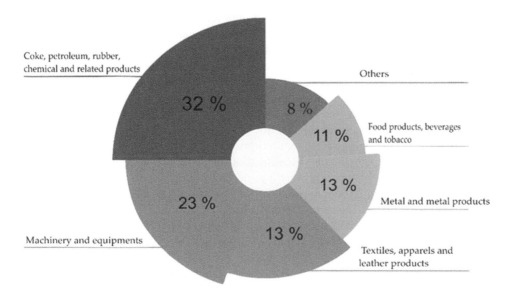

Coke, petroleum, rubber,
chemical and related products

32 %

8 %

11 %

Others

Food products, beverages
and tobacco

13 %

Metal and metal products

23 %

Machinery and equipments

13 %

Textiles, apparels and
leather products

Figure 20.2 Value generation from manufacturing, National Industrial Classification for Financial Year - 2019 (Ministry of Statistics and Programme Implementation [1]).

for process safety during the time of development phase [2]. Hazardous chemicals and procedures are used in the petrochemical industry, which can result in severe mishaps if human or equipment failure happens. Because most such enterprises are located near heavily populated urban or rural areas, particularly in developing countries, it is critical to investigate potential accidents and the damage they may do to implement preventive/mitigative steps before a tragedy occurs [3].

It is necessary to develop a plan to guarantee that industrial activities are carried out safely. The strategy's goal is to develop and maintain safe operations while keeping in mind the plant's design, operating circumstances, production needs, commercial requirements, and economic variables. In essence, the goal of process safety is to prevent uncontrollable occurrences in industrial processes [4]. The understanding of case studies and general knowledge about accidents is the first step toward enhancing the safety of industrial plants. According to a research on industrial safety, industry and the government believe that more and more information on accidents should be recorded and published year-wise [5].

In the process industries, there has been a growing demand for improvement in safety control during the last few decades. Traditionally, a considerable emphasis has been placed on identifying and controlling possible dangers posed by materials and conditions in process installations using techniques such as hazard and operability study (HAZOP), fault tree analysis (FTA), and layer of protection analysis (LOPA), as well as implementing appropriate controls. The vast majority of businesses rely on these approaches to maintain process safety. However, new scenarios tend to emerge regularly, resulting in mishaps that are beyond the control of current mostly technical analytical approaches [6].

Past accident analysis is one of the most effective and widely used techniques for understanding why accidents happen in the chemical process sector and the harm they create. Past accident analysis can be used to identify methods to prevent accidents or to lessen the impact of unavoidable accidents [7].

Risk management and process safety remain a major challenge for manufacturing and process industries; smart technologies have been applied over several decades to overcome these safety challenges, opportunities, and threats related to process safety management [8].

For safety management in GM, major obstacles like a large number of chemical hazards and environmental safety regulations are a prime concern. AI is used to overcome safety management issues. Identification of some technical challenges for the knowledge graph, incident early warning, decision-making, and risk-tracing can be accomplished by using AI techniques [9]. Smart technologies such as AI, big data analytics, and IoT are creating positive improvements for society in this Industry 4.0 era. Such technologies are deployed for environmental sustainability. Industries are relying on AI, big data, and IoT for better environmental practices [10].

As an emerging technology in knowledge representation, the knowledge graph is used for process safety in the chemical industries, semi-automatic construction framework including ontology, definition, data acquisition, import, and storage as in case of delayed coking process in the petrochemical industries for process safety [11].

The fourth industrial revolution has changed our society and livelihoods, including the digitalization of manufacturing industries. Several process industries have adopted Industry 4.0 technology and also identified various technical challenges related to process safety; knowledge extraction with scarce labels; knowledge-based reasoning, an accurate fusion of heterogeneous data from several sources; and effective learning for dynamic risk assessment and aided decision-making [12].

The rise of AI and its increasingly broad influence across various industries necessitates a review of its impact on achieving sustainable development goals to allow long-term growth. The rapid development of AI must be accompanied by the appropriate regulatory knowledge and supervision for AI-based technology. If this is not done, there may be breaches in transparency, safety, and ethical norms [13].

In the age of big data, intelligence, and Industry 4.0, intelligence plays an increasingly crucial role in the management and, more especially, decision-making in process safety; as a result, it has become a popular topic and is acknowledged as an important field. As a result, the term "safety intelligence" has been coined as a new safety concept. In the age of big data, intelligence, and Industry 4.0, safety intelligence is a new paradigm for safety science that strives to transform raw safety data and information into useful and actionable information for safety management. In the age of safety, it is seen as a necessary perspective for safety management. The big data, AI, and Industry 4.0 techologies play an increasingly crucial role in the management and, more especially for decision-making in process safety. As a result, they have become the popular topics and are acknowledged as important fields. The term "safety intelligence" has been coined as a new safety concept and a paradigm for safety science that strives to transform raw safety data and information into useful and actionable information for process safety management [14].

After several decades, process safety approaches have been applied in industry to make processes and systems safer and more efficient, thereby improving sustainability. The major goal is to keep people, equipment, and the environment safe [15].

20.3   FRAMEWORK OF PROCESS SAFETY FOR GM IN THE PROCESS INDUSTRY

Three core objectives of GM are: (a) Lessening of energy consumption; (b) Process safety monitoring; and (c) Evaluation of environmental imprint. This chapter proposes a framework of process safety for GM in the process industry, as shown in Figure 20.3.

AI is now regarded as one of the most innovative technologies, and it has had a profound impact on a variety of industries, including natural language processing, computer vision, and robotics. Furthermore, AI is often assumed to be essential for smart manufacturing. It plays an important role in attaining process safety management and enhancing efficiency by intelligent use of resources and energy consumption. Various AI techniques like knowledge graph, Bayesian network, and deep learning can be used for the design of alarming systems or early warning systems to avoid hazards in the process industry. The knowledge graph can be used for the integration of information (entity, relationship, and attribute); Bayesian network can be used for risk assessment and decision-making; and deep learning can be used for the design of early warning systems to prevent hazards in the process industry, as shown in Figure 20.3.

20.3.1   Artificial Intelligence

AI is a field of computer science and automation also known as machine intelligence. AI combines domain knowledge from computer science, automation, information engineering, mathematics, psychology, linguistics, and philosophy. The several subproblems of AI implementation in the context of process safety with a GM environment are shown in Figure 20.4.

The challenges to the execution of GM from the perspective of process safety can be divided into the following three groups based on their characteristics:

a.   Integration of information via Knowledge Graph
b.   Risk assessment and decision-making using Bayesian Network
c.   Early warning by using Deep Learning

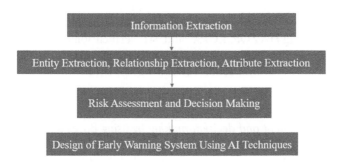

Figure 20.3 Framework for design of EWS.

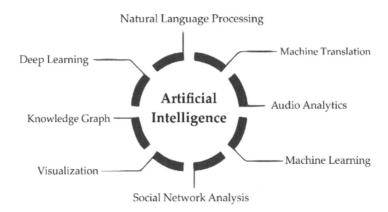

Figure 20.4 Several subproblems of AI in GM.

20.3.1.1   Integration of information via Knowledge Graph

In AI, the best tool for detecting connected data management is the knowledge graph as it aids in comprehending concepts and their relationships inside a system based on the type of semantic structure. This is ideal for online social networking, online financial systems, and online social security. This graph can provide reasoning and sophisticated rules based on the concept of deep learning. More specific knowledge is required in chemical engineering, safety, control, automation, and mechanical systems. The whole idea of the working structure of the graph is provided, and a knowledge graph can be created as shown in Figure 20.5 [16].

The key components of the knowledge graph are following:

- **Entity extraction:** The first stage is extracting information from text collections and automatically identifying named entities. A well-defined ontology is required to extract risk-related issues for the process industry. Some examples of data sources are maintenance datasheets and operation manuals.
- **Relationship extraction:** This is the second step in determining the entities' relationships. The relationship is usually represented by semantic data, tables, or diagrams. Process safety is usually established by process hazards safety analysis documentation in the petrochemical industry. Cause-and-effect relationships are documented using HAZOP, LOPA, and SIL verification documents.
- **Attribute extraction:** The third step is attribute extraction, which refers to process-related elements like relief valves, materials, and planned relief pressure, among other things.

20.3.1.2   Risk assessment and decision-making by using Bayesian Network

A directed acyclic graph is used to represent probabilistic correlations as well as conditional dependencies between variables by Bayesian networks which are probabilistic

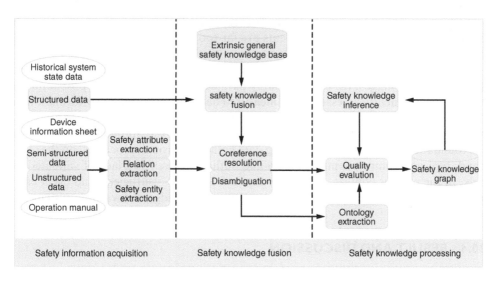

Figure 20.5 The technical architecture of knowledge graph [9].

graphical models [17]. Risk variables are predictably correlated to many types of ir-regularities and anomalies in the process sector via probabilistic correlations. Some major process parameters should be observed together with the reaction hazard in case of a runaway reaction in a coking furnace or a reactor for the delayed coking process which includes the temperature and flow of hot materials, heated materials along with the degree of coking within the furnace. Different parameters have dif-ferent risks of drifting from the norm. A Bayesian network can explain the intricate probability linkages between these parameters stated above and a runaway response along with other potential accidents and risk factors. A Bayesian network's parameters should be fine-tuned while keeping accuracy in mind to obtain optimal performance. As a result, prior knowledge-based parameters are commonly optimized using the principle of maximum entropy, while the estimate is frequently performed using the maximum likelihood technique. This methodology can search for an abnormal source to estimate whether a runaway reaction will occur in a coking furnace by thoroughly analyzing the data and can suggest emergency remedies such as distinct abnormalities.

20.3.1.3 Early warning by using Deep Learning

Deep learning is a subclass of ML which employs multi-layered neural networks to simulate the human brain's data interpretation function [18]. The possibility of an ac-cident in the process industry is often indicated by monitoring the volatility of process data. The changes in composition during a chemical reaction feed may lead to an in-creased heat generation within the reactors leading to a change of process parameter exceeding the safety limit and resulting in runaway reactions by a rapid temperature rise. A specific early warning function might be built if the links between the temper-ature/pressure rise in the upstream unit and potential explosive consequences in the downstream unit are established before the consequences. Unfortunately, there are just too many processes monitoring parameters, and the suggested linkages between parameter changes and possible threats are far too complicated for humans to comprehend [19].

Deep learning is typically used in this case to find patterns of possible accidents and, as a result, suggest the corresponding parameter adjustments. The probability of accidents is frequently assessed using measured data if comprehensive information on labeled process monitoring data, process equipment, and human operations is available. Deep learning and related risk detection and appraisal approaches are built on the foundation of labeled large data. Big data gathered through processes, on the other hand, typically comprises missing data and outliers, as well as a lack of validated labels. In the present context, accessible data is frequently insufficient, and event early warnings in industrial applications rely on expert knowledge and warning system.

20.4   RESULT AND DISCUSSION

Even though knowledge graphs have been applied in many specialized industries, this is still a novel method for the manufacturing process industry as complicated safety relations in the process industry knowledge graph still face many technical challenges. The challenges associated with knowledge graph for manufacturing process industry are as follows:

a.   **Knowledge acquisition with scarce labels:** When a knowledge graph is applied to the process industry, it may face numerous challenges because of the different chemical processes. A process safety application is a life-cycle action that requires the proper design of a process, equipment, automation, and human. If any of this fails, an event may occur. In the case of a chemical process's life cycle, data would come from various domains, making data collecting challenging. Furthermore, the most relevant data for process safety analysis is a real-time deviation of process monitoring with labels on aberrant situations and breakdowns.

b.   **Knowledge-based reasoning:** Process monitoring, irregular condition tracking, and outcome assessment should be swift and consistent due to the critical demand for risk reduction and safer operation in the process industry. Knowledge reasoning may expose some cause-and-effect links that humans are unaware of, offering additional knowledge for process safety analysis. State-of-the-art reasoning methods in knowledge graphs now achieve an accuracy of around 80%, though this level is inadequate for practical usage in process safety assessments. To meet the process industry's safety criteria, current knowledge reasoning approaches should be improved, or new techniques should be introduced.

c.   **Accurate fusion of heterogeneous data from several sources:** There are two types of process safety data: static and dynamic. Dynamic data refers to the status of a constantly changing process, whereas static data refers to process information and accompanying hazard assessment documents that do not change frequently. In reality, the data obtained is frequently ambiguous, making knowledge acquisition more difficult. There are two prospects. The first is related to data pre-processing, whereas the second is related to domain-based knowledge acquisition and data synthesis.

d. **Effective learning approaches for dynamic risk assessment and decision-making:** Knowledge graphs are an effective way to put together static knowledge and facts about the safety of petrochemical processes. In a chemical process, cause-and-effect interactions should ideally be maintained in the relationship among distinct entities and the rules or axioms that regulate them. Although knowledge graphs permit cause-and-effect assessment based on specific deviations, identifying the relevant departure might be challenging. To identify probable divergences from the usual condition, real-time data from the process monitoring system is essential. For unusual situations categorization, knowledge graphs and suitable ML methods are required. To do this, a multitude of data on process dependability, equipment failure mechanisms and their consequences, operational processes, and other topics is required. Unfortunately, high-quality data is rarely inadequate to enable algorithmic learning in reality.

20.5 CONCLUSION

The adoption and application of AI is already having a substantial impact on society, and it has the potential to play a significant role in the life-cycle process safety monitoring and risk management through hazard prevention in process industries in the GM environment. By combining imprecise and unpredictable data, AI may be further strengthened to produce real-time early warning systems. EWS was created with the use of AI to address the following issues. AI can be used to improve EWS to address the following problems.

- Real-time information on potential hazards
- Preventing the loss of lives and damage to the industry
- Predicting when and where hazards can occur
- Risk management and decision-making approaches related to hazards

With the use of AI, EWS can monitor and warn about possible hazards that proactive and preventive action can be taken to minimize the loss of lives and limit the damage to the process industry due to hazards.

REFERENCES

1. PWC, "India: A global manufacturing hub for chemicals and petrochemicals," no. March, 2021, [Online]. Available: https://www.pwc.in/assets/pdfs/publications/2021/india-a-global-manufacturing-hub-for-chemicals-and-petrochemicals.pdf.
2. F. Khan, S. J. Hashemi, N. Paltrinieri, P. Amyotte, V. Cozzani, and G. Reniers, "Dynamic risk management: a contemporary approach to process safety management," *Curr. Opin. Chem. Eng.*, vol. 14, pp. 9–17, 2016, doi: 10.1016/j.coche.2016.07.006.
3. F. I. Khan and S. A. Abbasi, "Rapid quantitative risk assessment of a petrochemical industry using a new software package MAXCRED," *J. Clean. Prod.*, vol. 6, no. 1, pp. 9–22, 1998, doi: 10.1016/s0959-6526(97)00045-0.
4. N. Gibson, "Process safety - A subject for scientific research," *Process Saf. Environ. Prot.*, vol. 77, no. 3, pp. 149–153, 1999, doi: 10.1205/095758299529965.

5. S. G. Balasubramanian and J. F. Louvar, "Study of major accidents and lessons learned," *Process Saf. Prog.*, vol. 21, no. 3, pp. 237–244, 2002, doi: 10.1002/prs.680210309.

6. B. Knegtering and H. J. Pasman, "Safety of the process industries in the 21st century: A changing need of process safety management for a changing industry," *J. Loss Prev. Process Ind.*, vol. 22, no. 2, pp. 162–168, 2009, doi: 10.1016/j.jlp.2008.11.005.

7. S. M. Tauseef, T. Abbasi, and S. A. Abbasi, "Development of a new chemical process-industry accident database to assist in past accident analysis," *J. Loss Prev. Process Ind.*, vol. 24, no. 4, pp. 426–431, 2011, doi: 10.1016/j.jlp.2011.03.005.

8. J. Lee, I. Cameron, and M. Hassall, "Improving process safety: What roles for Digitalization and Industry 4.0?," *Process Saf. Environ. Prot.*, vol. 132, pp. 325–339, 2019, doi: 10.1016/J.PSEP.2019.10.021.

9. S. Mao, B. Wang, Y. Tang, and F. Qian, "Opportunities and challenges of artificial intelligence for green manufacturing in the process industry," *Engineering*, vol. 5, no. 6, pp. 995–1002, 2019, doi: 10.1016/j.eng.2019.08.013.

10. A. K. Feroz, H. Zo, and A. Chiravuri, "Digital transformation and environmental sustainability: A review and research agenda," *Sustainability*, vol. 13, no. 3, pp. 1–20, 2021, doi: 10.3390/su13031530.

11. S. Mao, Y. Zhao, J. Chen, B. Wang, and Y. Tang, "Development of process safety knowledge graph: A case study on delayed coking process," *Comput. Chem. Eng.*, vol. 143, p. 107094, 2020, doi: 10.1016/j.compchemeng.2020.107094.

12. G. Dagnaw, "Artificial intelligence towards future industrial opportunities and challenges," *African Conf. Inf. Syst.*, 2020.

13. R. Vinuesa et al., "The role of artificial intelligence in achieving the sustainable development goals," *Nat. Commun.*, vol. 11, no. 1, pp. 1–10, 2020, doi: 10.1038/s41467-019-14108-y.

14. B. Wang, "Safety intelligence as an essential perspective for safety management in the era of Safety 4.0: From a theoretical to a practical framework," *Process Saf. Environ. Prot.*, vol. 148, pp. 189–199, 2021, doi: 10.1016/j.psep.2020.10.008.

15. A. Stolar and A. Friedl, "Process safety for sustainable applications," *Int. J. Reliab. Qual. Saf. Eng.*, vol. 28, no. 5, pp. 5–6, 2021, doi: 10.1142/S0218539321500339.

16. L. Ehrlinger and W. Wöß, "Towards a definition of knowledge graphs," *CEUR Workshop Proc.*, vol. 1695, Leipzig, Germany, 2016.

17. J. Zhu, Z. Ge, Z. Song, L. Zhou, and G. Chen, "Large-scale plant-wide process modeling and hierarchical monitoring: A distributed Bayesian network approach," *J. Process Control*, vol. 65, pp. 91–106, 2018, doi: 10.1016/j.jprocont.2017.08.011.

18. Y. Lecun, Y. Bengio, and G. Hinton, "Deep learning," *Nature*, vol. 521, no. 7553, pp. 436–444, 2015, doi: 10.1038/nature14539.

19. J. Zhu, Z. Ge, Z. Song, and F. Gao, "Review and big data perspectives on robust data mining approaches for industrial process modeling with outliers and missing data," *Annu. Rev. Control*, vol. 46, pp. 107–133, 2018, doi: 10.1016/j.arcontrol.2018.09.003.

Chapter 21

Sustainable recycling methods for different types of eco-friendly cutting fluids and their characteristics

An impetus for circular economy

S. Shaw, S. Awasthi, and Navriti Gupta

CONTENTS

21.1 INTRODUCTION

Machining processes are done on a large scale in industries, which calls for savings of both economics and materials. Efficiency is also of great importance, and it depends on choosing appropriate cutting tools to cutting fluid. Cutting fluid improves tool life and maintains surface roughness by cooling the tool and workpiece. Reduction in consumption of cutting fluids can be achieved by minimum quantity lubrication and minimum quantity cooling lubrication, which are examples for the application of cutting fluid which is economically more suitable than mist and wet application of cutting fluids.

From the perspective of cutting and machining conventional fluids served us very well, but it comes at a price of environmental damage and risk to human health. To counter such circumstances, researchers and techniques are being developed and studied for either proper disposal or reusability of these cutting fluids [1].

To achieve both economic and environmental betterment, eco-friendly cutting fluids are to be used which can be recycled after use [2]. Dry machining can also be

DOI: 10.1201/9781003293576-21

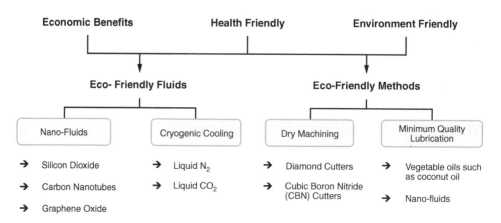

Figure 21.1 Circuit diagram for eco-friendly fluids and methods.

used, but it decreases the tool life and heat transfer doesn't occur which also affects the surface roughness. Some examples of eco-friendly oil are vegetable oil, semi-synthetic oil, liquid N_2, liquid CO_2 and so on. Cutting fluids are often dumped in the river bodies after their usage which leads to dangerous effects on the ecological cycle [3]. Hydraulic fluid and particulate matter present in used cutting fluids when dumped in water bodies cause pollution of mainstream water and promote bacterial growth; on the other hand, when incinerated and released into the atmosphere, they cause serious respiratory problems [4]. It is the need of the hour to use proper disposal methods of the cutting fluids so that it does not affect the environment at an alarming level [5].

The authors have reviewed different literatures regarding ecological cutting fluids to see if they provide enough lubrication and cooling in the cutting zone and their applications with techniques such as MQL (Minimum Quantity Lubrication). The discussion is firstly about various ecological cutting fluids with their physical, chemical and biological properties afterwards their applications in different machining processes such as turning and milling in various literatures have weighed in on the benefits of using MQL for vegetable oils and nanofluids [6].

There have been a number of experiments conducted with ecological cutting fluids such as experimenting with the turning operation on EN24 steel where they measured the quality of the fluid on the basis of MRR (Metal Removing Rate), surface roughness and machining force. They observed that at higher speeds, they achieved less surface roughness meaning smooth surfaces [7].

These investigations and experiments leverage the usage of eco-friendly fluids. Afterwards, at last, the recycling of the ecological cutting fluids and machining metal chip is discussed (Figure 21.1).

21.2   ECO-FRIENDLY CUTTING FLUID

The cutting fluids that can be recycled, or have a less polluting impact on the health of workers, environment, etc., are termed eco-friendly cutting fluids [8]. Such cutting fluids decrease occupational hazards and risks. Also, if less polluting / recycling plant-based cutting fluids are used, the chances of their entrance into the ecological food chains will be low as these cutting fluids wash away into drains and reach the water bodies [3]. Based on their origin, they can be broadly classified as follows:

1. Vegetable-based cutting fluids
2. Minimum Quality Lubrication (MQL)
3. Bio-oil cutting fluid
4. Cryogenic cutting fluids
5. Nanofluids

21.2.1 Vegetable-based cutting fluids

Vegetable-based fluids are one of the first choices due to health reasons and they also provide satisfactory lubricity which also ensures desirable surface roughness. Furthermore, economically, MQL can also be used for these oils. These oils are biodegradable and non-hazardous. The intermolecular interaction in these oils prevents them from raising their temperature and provides a high coefficient of viscosity. Vegetable oil is made of tri-glyceride structure which makes it a good fit for a lubricant. Long and polar acid chains contribute to the lubricating nature and strength of the fluid.

21.2.2 Minimum Quality Lubrication (MQL)

Minimum Quality Lubrication is a type of application which is concentrated only at the cutting zone, the pursuit of minimum usage of cutting fluid and maintaining efficiency. Contrary to wet and mist cooling, the usage of cutting fluid is very minimal which gives it an economic advantage and is less harmful. The principle of MQL is that a mist of air-fluid mixture is applied to the cutting zone through the machine tool spindle [5]. It promotes evaporative and convective heat transfer due to which it is effective in cooling as compared to wet cooling [1].

In MQL, not only is it required to lubricate the surface but also to flush out the chips from the machining zone which is necessary for sanitation as well as it doesn't get accumulated there resulting in interference of the machining process. Xavior and Adithan [9] studied tool wear and surface roughness of AISI 304 by performing turning operation by taking inputs such as cutting speed, feed rate and depth of cut. Findings indicated that coconut oil is comparable to traditional cutting fluids for reducing tool wear and surface roughness.

Surface roughness and tool wear are results which are used for the assessment of cutting fluids in turning operations which also do not only depend on cutting fluid but feed and cutting velocity also. Vegetable oils can also be used with MQL to achieve better performance with an economical advantage.

21.2.3 Bio-oil cutting fluid

The commercial market for bio-oil in various applications is large and bio-oil can also prove to be cheap if used in industrial applications such as machining. A wide variety

Table 21.1   Examples of strains [10]

Bacteria	Yeast
Bacillus	Saccharomyces
Brevis	Cerevisiae
Subtilis	Candida lipolytica
Pseudomonas	Mortierella
Fluorescens	Isabellina

of bio-oil can be selected with their qualities such as non-toxicity and easy availability. The protein or the oil that is produced by the microorganisms can be used because of its lubricating properties; thus, they can be used as cutting fluids.

After identifying the correct microorganism and testing it, cultivation plants for bio-based oil can be established. Tests could be conducted for pH value, viscosity and specific gravity in the early stages of development [10].

Some examples of strains (Table 21.1) which could be applicable for cutting fluids have been identified and a few of them could be capable for the production of bio-based cutting fluids.

Teti and Segreto [11] conducted an experiment of turning operation with microbial, conventional and dry cutting fluid. The microalgal species Spirulina arthrospira-based cutting fluid demonstrated better performance against tool wear than conventional cutting fluids and dry cutting for AISI 1045 steel. Tool wear was evaluated with flank wear at fixed intervals of time. The surface roughness values with bio-oil showed better results than dry cutting fluids and were comparable to conventional cutting fluids.

Commercial water-based metal working fluid and microbial-based metal working fluid were used. The operation was evaluated upon the worn area, cutting forces, tool wear and surface finish. The microbial-based metal working fluid leads to improved surface finish and reduced tool wear in different concentrations. The influence of cell size in lubrication in this experiment was indicated.

21.2.4   Cryogenic cutting fluids

Cryogenic fluids have considerably lower temperatures than other cutting fluids which are beneficial for the cooling action of the cutting zone; and because of their gaseous state, they tend to reach more space, thus increasing their efficiency. But sometimes due to their low temperature, it also increases the chance of the workpiece getting its physical properties changed which is undesirable. The selection of the cryogenics could be done by selecting the gas which is present in the environment plenty enough and doesn't react in any way with machinery and is also less harmful to the environment. For example, liquid N_2 and CO_2. Cryogenic fluids could be used if the temperature has to be kept low and regulated. Jerold and Kumar [12] investigated turning operation on AISI 1045 steel using cryogenic carbon dioxide in both dry and wet conditions using conventional cutting fluid by varying feed rate and cutting velocity, by judging the results with temperature in the cutting zone, chip thickness and surface roughness. In high-speed machining, the conventional cutting fluids had a lower temperature. The cutting forces in the cryogenic conditions were less than in dry and wet conditions due

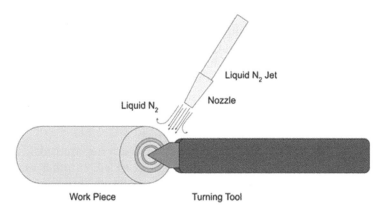

Figure 21.2 Simple working diagram of liquid N_2 turning operation.

to the formation of a lubricating layer which reduces friction. The surface roughness was also lower than in wet and dry conditions. Ravi and Gurusamy [13] conducted an experiment of turning operation on AISI p20 steel with liquid nitrogen cooling in a dry and wet environment using traditional fluid. The wear and tear in the cryogenic environment were less than in dry and wet conditions resulting in better tool life and reducing the chip curl than in other conditions.

Liquid nitrogen (Figure 21.2) can be utilized by using a nozzle and directing it towards the cutting zone in the case of a turning operation. The pressure could be adjusted as per the requirement and the distance between nozzle and cutting zone is to be adjusted so that effective cooling could be provided.

21.2.5 Nanofluids

Heat transfer is one of the most important qualities for cutting fluids. Nanofluids have promising heat transfer capabilities which maintain the temperature of the cutting zone. Size effects and Brownian motion are methods of energy transfer in these fluids. These nanofluids are mainly prepared by two methods which are as follows:

i.   **Single-step Method:** This method combines the preparation and synthesis of nano-fluids by physical vapour deposition and liquid chemicals.
ii. **Two-step Method:** The nanofluids are developed by dispersing the nanoparticles in the base fluid. Nanoparticles are produced by inert gas condensation and the chemical deposition method [14]. Some examples of widely used nanoparticles are silicon dioxide (SiO_2), carbon nanotubes and graphene oxide.

Kilincarslan and Cetin [15] conducted an experiment of milling operation on AA7075-T6, specifically surface milling, using three varieties of cutting fluid by mixing in certain proportions of ethylene glycol, nano-silver and boric acid, the first one consisting of ethylene glycol and nano-silver (a), the second one consisting of ethylene glycol and boric acid (b) and the third one consisting of ethylene glycol, nano-silver and boric acid (c). Cutting fluid (c) had no significant wear on the tooltip due to lubrication and cooling achieved by it. Cutting fluid (a) had no oil film; so, the surface was not adhered to the particles which is why the chips had irregular shapes. Cutting fluid (b) despite having a protective oil film, chips had irregular structures due to poor cooling performance. Cutting fluid (c) had both the oil film and cooling capabilities of nano-silver; so, the shapes of chips were both regular and smooth.

21.3 RECYCLING METHODS

Conventional cutting fluids having emulsion of oil and water are widely used in machining processes. Oil in water emulsions can be recycled. In other words, oil and water can be separated and filtered and can be of use for any kind of purpose. These emulsions are created with stability using surfactants and other techniques; if this stability decreases, their separation can be done using gravimetric methods. Electro-chemical cells can also be established to separate oil and water.

In Figure 21.3, various cutting fluids recycling methods have been shown. The cutting fluid can be recycled depending upon the type of undesirable substances found in the used cutting fluid after the machining process is over. Waste cutting fluid can be filtered by centrifugal processes, magnetic separation and membrane separation [14]. Recycle plants can also be established in which fine chips and other debris can be crushed using a crusher. After crushing magnetic and centrifugal-based separation, filtration and sterilization are the next processes for the waste fluid. Fluids can also be heat treated to reduce the viscosity so that it is hard for fine particles to flow easily and then sterilizing the fluid for the safety of workers [16].

Bio-based cutting fluids which have microorganisms could be used as an energy source and the fluid after use can be used to manufacture high-end products by recycling and using it as raw material for biofuels [10]. Cutting fluid can also be recycled or treated using bio-degradation. Jagadevan et al. [17] showed that treatment of metalworking fluids with ozone reduces toxic and refractory components into biodegradable intermediates so that metalworking fluids could be disposed of properly.

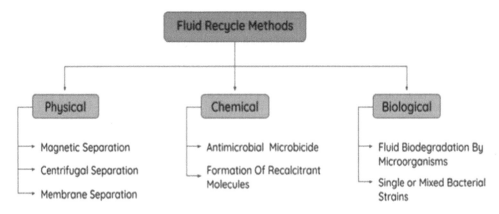

Figure 21.3 Different fluid recycling methods.

21.4 CONCLUSIONS

Ecological cutting fluids undoubtedly provide a safe environment to work, and with proper combination of fluids and methods implemented, we will get the same performance as conventional techniques and could even surpass them in future. With this, the scale of machining would be both close and much safer. Using techniques such as MQL-assisted machining operation with ecological cutting fluids could also prove less expensive. Cryogenic-based fluids and nanofluids could be used on a commercial scale because of their larger availability. To reduce the toll on the environment, these fluids could be recycled and put to use in industries once again. Physical, chemical and biological ways can be used for recycling of cutting fluids, each consisting of various techniques and processes. In practice, the fluid is either safely dumped or chemically recycled after use. Biological recycling methods may be applied in large plants and factories where recyclability can be easily administered and carried out. The mineral-based cutting fluids should be used only if necessary because these fluids can also pose negative effects if the machining involves using ceramic tools. The machining where interrupted cutting is involved with carbide tools such as milling where wear is predominantly of thermal origin which is produced by cyclic variation of temperature, cutting fluid will only hurt the process by increasing temperature variation [18].

21.5 FUTURE SCOPE

Ecological fluids seem promising to replace conventional cutting fluids in both conventional and unconventional types of machining processes [19]. These fluids could prove to be cost-effective using MQL (Minimum lubrication quantity) if the performance does not downgrade. A hybrid model of cryogenic machining and nanofluids can be deployed with or without MQL. The cooling could be done in phases one after the other using very small quantities of each fluid. For example, in the first phase, cryogenic cooling could be used to reduce the temperature; and in the later phase, the nanofluid can be used in little amounts. The cooling and lubrication for tools and workpieces could be separately done. Because of their different properties, they may require different cutting fluids but final consumption may be less than only using one cutting fluid. This could be beneficial, especially regarding ecological cutting fluids, because of their varying capabilities. Several types of hybrid cutting fluids containing environmental friendly components are under development that will tackle the challenges faced by pure ecological cutting fluids in the future.

21.6 RESULTS AND DISCUSSIONS

There has been extensive experimentation and investigation done with eco-friendly cutting fluids which are seemingly hopeful to get the outputs such as good surface finish, increased tool life and lowered cutting zone temperature. Recycling of some ecological cutting fluids as well as the metal chips that are generated during machining also prevents pollution.

Cryogenic fluids, on the other hand, just evaporate in the air and they could be generated via a plant dedicated to generating cryogenic fluids such as nitrogen and oxygen, but the other fluids need to be recycled. The literature so far has shown using different types of cutting fluids and also varying concentrations and mixing them with each other such as the case of vegetable-based cutting fluids also using the mist of their fluids to achieve cooling and lubrication.

The cutting fluids should be used only if necessary. The cutting fluids can be used in combination to reduce corrosion and to achieve advantages of different fluids.

REFERENCES

1. Xavior, A., & Adithan, M. (2009). Determining the influence of cutting fluids on tool wear and surface roughness during turning of AISI 304 austenitic stainless steel. *Journal of Materials Processing Technology*, 209, 900–909.
2. Kiran, Tomar, H., Gupta, N. (2021). Sustainability Concerns of Non-conventional Machining Processes—An Exhaustive Review. In: Agrawal, R., Jain, J.K., Yadav, V.S., Manupati, V.K., Varela, L. (eds) *Recent Advances in Smart Manufacturing and Materials*. Lecture Notes in Mechanical Engineering. Springer, Singapore. https://doi.org/10.1007/978-981-16-3033-0_25.
3. Soković, M., & Mijanović, K. (2001). Ecological aspects of the cutting fluids and its influence on quantifiable parameters of the cutting processes. *Journal of Materials Processing Technology*, 109(1–2), 181–189.
4. Haider, J., & Hashmi, M. S. J. (2014). Health and environmental impacts in metal machining processes. *Comprehensive Materials Processing*, 8, 7–33.
5. Debnath, S., Reddy, M. M., & Yi, Q. S. (2016). Influence of cutting fluid conditions and cutting parameters on surface roughness and tool wear in turning process using Taguchi method. *Measurement*, 78, 111–119.
6. Sen, B., Mia, M., Krolczyk, G. M., Mandal, U. K., & Mondal, S. P. (2019). Eco-friendly cutting fluids in minimum quantity lubrication assisted machining: A review on the perception of sustainable manufacturing. *International Journal of Precision Engineering and Manufacturing-Green Technology*, 8(1), 249–280. https://doi.org/10.1007/s40684-019-00158-6.
7. Kumar, P., & Ravi, S. (2021). Investigation on effects of vegetable-based cutting fluids in turning operation of "EN 24 Steel." *Materials Today: Proceedings*, 39, 95–99. https://doi.org/10.1016/j.matpr.2020.06.315.
8. Carvalho, D. O. A., da Silva, L. R. R., Sopchenski, L. et al. (2019). Performance evaluation of vegetable-based cutting fluids in turning of AISI 1050 steel. *International Journal of Advanced Manufacturing Technology*, 103, 1603–1619. https://doi.org/10.1007/s00170-019-03636-y.
9. Xavior, M. A., & Adithan, M. (2010). Evaluating the performance of cutting fluids in machining of AISI 304 austenitic stainless steel. *International Journal of Machining and Machinability of Materials*, 7(3/4), 244–259.
10. D'Addona, D. M., Teti, R., & Conte, S. (2020). Feasibility study of using microorganisms as lubricant component in cutting fluids. *13th CIRP Conference on Intelligent Computation in Manufacturing Engineering*, CIRP ICME'19.
11. Teti, R., D'Addona, D. M., & Segreto, T. (2021). Microbial-based cutting fluids as bio-integration manufacturing solution for green and sustainable machining. *CIRP Journal of Manufacturing Science and Technology*, 32, 16–25.
12. Jerold, D. M. & Kumar, M. P. (2011). Experimental investigation of turning AISI 1045 steel using cryogenic carbon dioxide as the cutting fluid. *Journal of Manufacturing Processes*, 13, 113–119.

13. Ravi, S., & Gurusamy, P. (2020). Cryogenic machining of AISI p20 steel under liquid nitrogen cooling. *Materials Today: Proceedings*, 37, 806–809.

14. Li, Y., Zhou, J., Tung, S., Schneider, E., & Xi, S. (2009). A review on development of nanofluid preparation and characterization. *Powder Technology*, 196, 89–101.

15. Kilincarslan, S. K., & Cetin, M. (2020). Improvement of the milling process performance by using cutting fluids prepared with nano-silver and boric acid. *Journal of Manufacturing Processes*, 56, 707–717.

16. Baradie, M. E. (1996). Cutting fluids: Part II. Recycling and clean machining. *Journal of Materials Processing Technology*, 56, 798–806.

17. Jagadevan, S., Graham, N. J., & Thompson, I. P. (2012). Treatment of waste metalworking fluid by a hybrid ozone-biological process. *Journal of Hazardous Materials*, 244, 394–402.

18. Sales, W. F., Diniz, A. E., & Machado, Á. R. (2001). St52 E355 E410 E470 20mnv6 C45e Ck45 St44 Cold Rolled and Cold Drawn Seamless Steel Precision Honed Hydraulic Cylinder Tube. *Journal of the Brazilian Society of Mechanical Sciences, 23*(2), 227–240.

19. Kumar, L., & Gupta, N. (2021). The effect of different parameters on MRR, surface finish while EDM machining titanium alloys: A review study. *Recent Advances in Smart Manufacturing and Materials*, 321–326.

Chapter 22

Sustainable automobiles

Major obstacles on the path of electrifying mobility in India, existing barriers and challenges

A. Arora and Navriti Gupta

CONTENTS

22.1 INTRODUCTION

The concept of sustainability has led the transportation and automotive industry to include environmental concerns in any innovation and product launch. The decreasing reserves of petroleum and other fossil fuels have led to the implementation of sustainability. EVs are made up of lightweight composites and alloys, which further decrease the demand for conventional metals and thereby reduce the impact of mining on the environment. However, EVs are still evolving and there are still limitations and doubts regarding battery capacities, impacts of materials and methodologies used [1]. Here, we review certain barriers to EVs in India, reviewing 11 major concerns and their effects. Firstly, we point out ongoing steps and how the Indian market is adapting to this shift. Then after a brief review of an EV, we present the major barriers and briefly

DOI: 10.1201/9781003293576-22

discuss each. This review study ends with reviewed numbers, a discussion of future plans and the presentation of conclusive statements.

India has recognized the urgency to develop cleaner and more effective transportation solutions to reduce dependence on international oil, control CO_2 emissions and with the vision to establish itself as a manufacturing hub, it seeks to switch to electricity. The Indian government has set itself an ambitious target of complete electrification in the country by 2030, for which the NITI Aayog (think tank) has been outlining the long-term strategy [2]. According to the Automotive Mission Plan 2016–2026, with the introduction of EVs, the Indian automotive sector is expected to become a huge engine of job creation for the country with around 6.5 crore jobs to create over the decade. Although the EV market share is currently less than 1%, the country's ambitious target of global EV30 @ 30 initiative, which sets a goal of 30% new EV sales by 2030, expects to add around 2.5 crore 2-wheeled vehicles, 29 lakh 3-wheeled vehicles and 54 lakh 4-wheeled vehicles [3]. Presently, around 23 Original Equipment Manufacturer (OEMs) offer EV models which are circulating in the Indian market. In India, currently, these vehicles are coming across various assessments like life cycle assessment, high charging time, driving range anxiety and insufficient infrastructure. Although technological advancements, incentives for the purchase and development of public charging stations are looking to accelerate the growth of EVs in India.

22.2   LITERATURE REVIEW

The study of dynamic travel schedule and charging profile was studied in a research study by Brady and Mahony [4]. They carried out a simulation of the propulsion system of the EVs. In the experimental research, it was found that on increasing parking time distribution, the overall accuracy of the system increased. In another analytical study by Morrissey et al. [5], it was found that EV customers prefer to charge their car at home in the evening time during peak demand electricity. In yet another study on EVs, Foley et al. [6] studied the impact of peak charging and peak-off charging while EV charging in Ireland. In the research, it was found that peak charging is more harmful battery charging as compared to off-peak charging. In yet another study by Doucette and Mc Culloch, the carbon dioxide emissions from the Battery Electric Vehicle (BEV) and the Plug in Hybrid Vehicles (PHEV) in comparison to CO_2 emissions from Ford Focus. In another analytical research, Steinhilber et al. [7] found out how to introduce new technology using different methods and strategies by finding out key barriers. In another research, Hai et al. [8] using a trip segment partitioning algorithm introduced a driving pattern recognition methodology in order to evaluate the driving range of the EVs. In yet another study, Hayes John et al. [9] build a vehicular model to investigate different driving conditions. In yet another research, Salah et al. [10] studied the EVs charging impact. They concluded that risks of overloading increase with higher penetration levels and dynamic tariff increases. The effect of various charging methods on the National Grid and their storage details is given in another research [11]. The maximum transmissible torque method is determined in another research [12] for increasing the antiskid execution of the torque control framework and improving the stability of the EVs. In yet another review study by Lu et al. [13], Li-ion battery management for the case of EVs was discussed. The issues associated with the EVs and their battery such as voltage optimum for the battery cell, the battery state estimation, equalization of battery and uniformity of the battery system and its fault analysis can form the basis of research for the design of the advanced EVs, and its equally important battery management system in a review study [14] energy management system and the simulation and 3D modeling methods were thoroughly studied.

22.3   GROWTH AND STRATEGY

Basically, the traction motor receives the power of a rechargeable battery (which can be charged externally) through a power controller. The motor uses this power and voltage supply received from the controller to propel the vehicle. While braking, this motor acts as a generator and recharges the battery pack improving the overall efficiency of the drivetrain. Sensors are attached to the accelerator and brake pedals which provide signals to the controller [15]. For rapid growth in this sector, it is imperative that India swiftly develops a strong Research and Development (R&D) capacity to develop and commercialize EV subsystems according to its needs. The government would play a major role in aiding and promoting R&D through grants and facilities. The industry needs to focus on the grand challenge of the development of vehicles according to Indian standards of costs, roads and usage. Immediate short term R&D goals are essential for more efficient and cost-effective motors and controllers, faster and more robust EV chargers, steering and stability systems, power brakes and regenerative braking systems, efficient electrical air-conditioners, vehicle safety and control management systems and in-vehicle communication protocols. All these R&D proposals would help to establish target costs and manufacturing methods for EV subsystems in India. Further intense R&D is required to neutralize the impact of large scale charging infrastructure on the existing electric grid and the development of cleaner energy. Also, India should prepare itself for new approaches to mobility and promote long-term R&D in all aspects of modern EV technologies like hydrogen fuel cells, autonomous vehicles, new battery-chemistry withstanding higher temperatures and higher energy densities, distributed lightweight motors, heavy electric trucks, Grid to vehicle (G2V) and Vehicle to Grid (V2G) connections and smart charging [16]. Given certain constraints (cost sensitivity) and opportunities (vast population), India's EV strategy should evolve with major emphasis on:

1. Improving the energy efficiencies and cost-effectiveness of EVs
2. Developing optional battery replacement for charging and building stations and swap points
3. Creating an efficient battery manufacturing ecosystem, from resources to batteries and proper recycling
4. Generating massive demand, especially with electric public transport [17]

22.4   BARRIERS TO ELECTRIC VEHICLES (EVs)
IN THE INDIAN MARKET

Different perspectives such as market, policies, technical barriers and lack of infrastructure need to be addressed before proceeding with the EV revolution in India.

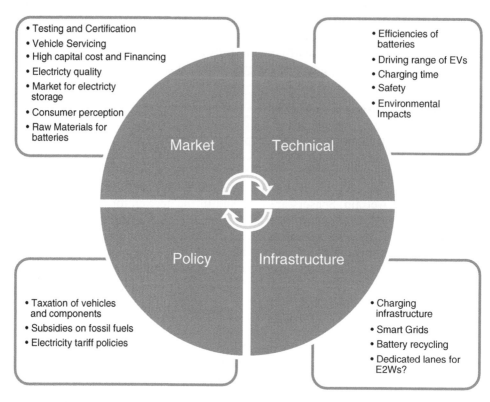

Figure 22.1 Barriers to EVs [18].

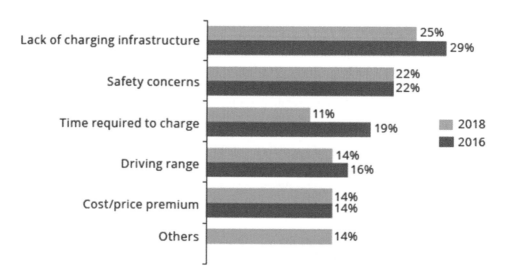

Source: 2020 Deloitte GACS (n=3,022)

Figure 22.2 Indian consumer concerns regarding EVs [19].

Further major consumer concerns should be accounted for while developing the EV strategy for India (Figures 22.1 and 22.2).

22.4.1 Skill gap

For proper and regular maintenance of the car, an experienced technician with expertise in EV should be able to troubleshoot and repair the EV. However, this is a problem due to the lack of adequate knowledge of EV subsystems among conventional technicians. India is the youngest nation on the planet with an average age of 29 years (compared to Chinese 37), but the huge gap in the quality of skills is worrying and the available skilled talents are fragmented. It is necessary to establish an adequate ecosystem of modern technologies and techniques in industry and education. It requires effective coordination between the different institutions and cannot be done alone. The shift to EVs is fueling the demand for specialists in mechatronics engineering, battery technologies, lightweight composites and motor control systems which are currently scarce in India. However, alongside leading organizations such as SIAM and ACMA, some major OEMs and suppliers are promoting modern skills development and training a new generation of skilled young people in India [20].

22.4.2 Cost constraint

The most important factor for the relatively slower penetration of EVs in the Indian market is the price constraint. In India, less emphasis is given to convenience when buying a vehicle compared to the western world. People mainly avoid purchasing EVs due to current purchasing costs which result in a weaker supply chain and associated price increases. The high costs of the vehicle are mainly attributed to their expensive battery packs. They can even represent 40–50% of the cost of the vehicle and also need to be replaced more than once in its life.

Although you factor in the cost incurred over the lifetime or the cost of operation, an EV is much more economical. This is mainly due to less maintenance cost over its lifetime and the relatively cheaper electricity. In addition, the continuous improvements and advancements in battery design and chemistries will certainly reduce costs (which have been very significant over the past decade) and, ultimately, make EVs affordable to the masses. They are expected to have a similar price to conventional vehicles with the same specifications by 2025 [16]. Currently, manufacturers will have to compromise and come up with mid-range EVs to keep vehicles affordable for the masses. Although this, in turn, requires more frequent and faster-charging installations which is another problem in India [21].

22.4.3 Consumer perception

Although the range of EV variants in the markets keeps expanding, the decision of the customer to purchase an EV is still uncertain. Studies indicate that the public, however, is considerate of environmental factors and is in agreement that EVs may be

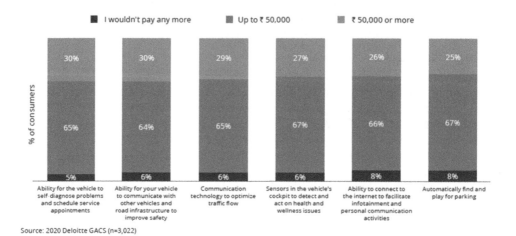

Source: 2020 Deloitte GACS (n=3,022)

Figure 22.3 Indian customers' willingness to pay for various connected technologies [19].

a better alternative but still a good percentage of the Indian population does not trust the technology of EVs and this further discourages them. The majority of consumers in India believe that the assisting facilities and improper charging station setup are the prime reasons for them not choosing an EV. Lack of knowledge about EVs among the Indians is also a great concern because a large piece of the population has no proper idea or simply just heard of the EVs. Further social considerations need to be taken into account as studies show them to be pronounced among Indians [22]. Another important factor for swift adoption in the market would be to symbolize a social status and branding associated with EVs.

Young people are more inclined to adopt EVs more quickly. In addition, customers with adequate academic training respond positively to the demand for electric mobility. However, the region of residence generally does not have a large influence on EV adoption, but the customer's employment and income do impact the demand for EVs [23]. Therefore, consumers should be informed of the benefits through various marketing channels. Lack of awareness of schemes, financial incentives and safety advancements would have a major effect (Figure 22.3).

22.4.4   Raw materials

Major inputs for EV production are conventional vehicle materials such as aluminum, copper and steel as well as some battery constituents. Lithium-manganese-cobalt oxide (NMC), lithium-cobalt-aluminum oxide (LCA), lithium iron phosphate (LFP) or, to a limited extent, lithium manganese oxide (LMO) are used at cathode and graphite at anode in major Li-ion batteries. Mass manufacturing in India would need a strong supply network with lithium, cobalt, nickel and graphite supply being extremely smooth. The global distribution and manufacturing efficiency of rich cobalt resources make it costly and largely unavailable. The largest accessible lithium reserves mainly exist around the salt lakes of Chile, Argentina and Bolivia and around Australasia. The dominant cell chemistry in India would rely on the specific energy

Table 22.1 Capacities and consumption in 2020 for major EV segments [3]

Segment	Net capacity (in kWh)	Consumption (in kWh/km)
Electric bike	1.2	0.016
2-wheeler	2.2	0.025
Rickshaw	3.7	0.043
Personal car	40	0.157
Heavy trucks	27	1.500

density requirements and access to raw resources across the country. To ensure a consistent supply of important inputs to the Indian EV sector, the government has administered a collaboration of three national companies called Khanij Bidesh India Ltd. (KABIL), which is now a partner of an Argentine govt.'s mining company which has been recently looking after lithium and is also planning for ties with Bolivia [3]. So, by encouraging batteries containing lesser cobalt and providing incentives for crucial items, India might be able to cut down dependence on imports and develop required technical strategies.

22.4.5   Battery life

EV battery packs are ideally designed to work all vehicle's life, although in practice they eventually degrade. The cost of the new pack is currently never sufficiently explained by companies; but if such a situation arises outside the warranty period, it usually means a big cost for such replacement. Currently, most major manufacturers offer an 8 year / 100,000-mile warranty for their batteries [24] (Table 22.1).

Industry regulated NMC (cathode) / graphite (anode) as the standard and started building the battery ecosystem. This mainly includes the production of battery packs using the cells, the production of cells and the attachment of various useful materials and chemicals. India has dominated the production of engineered pack cells with proper heat flow designing and quality checks to ensure efficient operation in tropical and subtropical regions, also proper structural and mechanical designing with checks to make sure every cell is exposed to optimal pressure conditions, finally battery management system; to ensure a balanced charge and discharge of all cells in the pack and ensure none of them has runaway issues [17].

Battery constituents such as lithium, nickel, cobalt, manganese and titanium don't only financially strain the supply network but are also harmful to the surroundings when they are scraped off. India, which already has these resources, has started reusing and recycling existing batteries and is dominating the recovery of almost all of these important items from waste packs in an eco-friendly way [17]. The economics of recycling or reusing an existing battery is usually firmly anchored in favor of reuse. The major parts are: (a) the reconditioning cost to reuse the battery pack; and (b) the credit accumulated by recycling the battery. The price of reuse is lower than the cost of restructuring, and the recycling credit should therefore be financially preferred. The main recycling methods used for lithium-ion batteries include pyro-metallurgical recovery, physical separation, recovery of hydrometallurgical metals, direct methods recycling and recovery of biological metals [25].

22.4.6   Driving range

A range is considered to be the main barrier of the EV in general, as even modern EVs have a significantly lower driving range than a comparable conventional vehicle. The net area that an EV may cover when fully charged is the main disadvantage to absorbing the EV in the global market and in India. Battery size and capacity, cooling requirements, gross vehicle weight, power-to-weight ratio, vehicle design and speed, road conditions and vehicle fuel efficiency (amount of energy required per km), all contribute to the range a vehicle can cover before needing to be recharged or changed. The resolution of the autonomy problems associated with EVs is crucial for the development of the ecosystem [16].

The Indian vehicle market is dominated by two-wheelers, the existing ranges announced by major variants need to rise two-folds to a minimum of 200 km to compete in the Indian market. Most of the vehicles presently provide a range of under 250 km per recharge. Although a number of the recent advanced models can give up to 400 km, then the cost rises with increase in mileage . So, in shopping for an EV, the motive force would have to be compelled to set up travel fastidiously and will have no choice for a longer trip. So, presently, the driving range of EVs seems like a major barrier to growth [24].

22.4.7   Duration of charging

The time it takes to recharge a vehicle is closely concerning the problem of range anxiety. Standard charging from a 7 kW point, the energy unit might need 8–10 hours to totally recharge. Such time requirement varies according to battery capacity with larger packs taking more time. Secondly, the rate is influenced by charging point specifications as the more rate of the station, the lesser would be the net duration. Presently, fast chargers are being employed to recharge EVs minimizing the charge time. All modern production EVs are created so as to handle such a rate; therefore, the battery is currently charged at the most optimum rate that they will handle without failure. However, the rate may vary according to temperature and in cold conditions charging may be slow. In DC quick charging, AC to DC conversion happens within the charging station itself. It is the quickest charging arrangement because it supplies immense power and recharges the vehicles at a very fast pace. It would just take 30 minutes to get ready for a 145 km range [24]. It ought to be noted the fast charging infrastructure entails a huge impact on the existing electrical grid owing to its sharp demand, i.e., usually over thrice of a median home [26].

22.4.8   Safety regulations

Battery packs must be able to handle extreme conditions such as rapid temperature hikes, circuit failures, heating, vibrational noises, collisions or water immersion. Design specifications must be optimized to incorporate major safety subsystems such as Battery Management Protection (BMS) protection, circuit and voltage isolation and collision avoidance modules. The battery, which poses a serious threat, must be designed and installed with extreme caution. During reconditioning and recycling, the battery packs are dismantled at the single module stage. This requires intensive training with extreme voltage conditions and miscellaneous tools to avoid short-circuiting or electrocution as such conditions may cause thermal instability and give rise to harmful by-products like the HF25 gas that on combining with its subordinates starts filling up inside and eventually leads to the explosion of the battery [25].

22.4.9 Environmental factors

The rising stress to manage the quantity of greenhouse emissions and raise air quality standards has compelled nations to take concrete steps. Such environmental factors critically have an effect on the promotion of adopting cleaner quality alternatives akin to EVs. The environmental effects of the inner combustion engine vehicles are fatal [23]. It may be a general notion that the EVs don't affect the environment; however, extraction mining and brines in desert regions add to it [15]. It takes roughly 750 heaps of intensified and rich brine to produce just a ton of Li (Lithium). The process and extraction methodologies of huge amounts of raw materials have environmental impacts. Such a process affects the geological formation of the area. In Salar de Atacama, a major location for Li cell production in Chile, more than half of the area's water resources are being used up by these activities. This severely affects the farmers in the area who are being forced to import water from different regions [25].

22.4.10 Government policies

Government policy lays the foundation for sustainable production in a country. The Indian government has gone to great lengths to formulate effective policies that will facilitate the introduction of EVs in the market by announcing various subsidies and incentives on purchase. State administrations are working on minimizing tariffs on charging. Facilities like reserved parking and new insurance policies are accelerating the growth in the country [23]. To further accelerate the electric car revolution, the Indian government is hoping to even subsidize the complete charging infrastructure for electric cars in India. The Department of Energy made it very clear that charging stations for EVs in India do not require a license. The government should rapidly reduce the GST rate for lithium battery packs, provide benefits to lure as many EV buyers as they can and focus on shifting the public transport sector to EVs [24].

The National Electric Mobility Mission Plan (NEMMP) 2020 was declared by the Indian government to improve the country's energy security by reducing dependence on imports, mitigating the adverse effects of fossil fuel vehicles on the environment by promoting cleaner mobility solutions and developing national production capacity. The Indian government has also brought about planning for Faster Adoption and Production of Electric Vehicles (FAME II) to boost the rapid adoption of EVs in India. In February 2019, the government authorized approximately Rs. 10,000 crore for the implementation of FAME II from 1 April 2019 for a period of 3 years.

NITI Aayog's 2017 Transformative Mobility Report laid down a roadmap for the deployment of BEVs across the country. It is believed that if India adopts transformative solutions of collaborative and connected mobility, like all public vehicles electric and half of the personal vehicles electric, it can fulfill its aim of being fully electric by 2030. The Indian government is administering around 100 GW of power generation via solar by 2022, which could have a direct impact on the availability and reliability of energy [24].

22.4.11   Charging infrastructure

A huge charging setup is essential to accommodate increasing EVs in India, resulting in increased demand for electricity. By comparing the potential impact of such vehicles in the grid and the US market with the Indian counterpart, we see that, currently, India won't even be able to manage 10% of its vehicle fleet as electric [26].

There is no doubt that the charging infrastructure is supposed to play an important part in the implementation of EVs in the country; and with improper planning and without a timed schedule, it could turn into a major obstacle for mass adoption. The setup consists of all amenities for efficient movement of electricity from the grid to the vehicle. It is evident that due to the shortage of adequate charging stations, the sales figures are low. The government must start implementing the necessary measures to quickly install the infrastructure for charging EVs. The first step should encourage players in the international market to focus on potential positions; instead of directly involving government entities, private companies should be outsourced to keep the workflow smooth.

India needs to use the available resources and network wisely. This can be achieved through coordinated charging and energy-saving principles. Increasing supply through the installation of power stations is a time-consuming process. That is why we have to switch to smart grids. Smart grids create a bi-directional connection through which power and data can move smoothly across companies and their users. The main goal of the smart grid would be optimizing the electricity demand by installing a smart energy management system in every home and workplace that will help control electricity usage more efficiently. Devices and household appliances will adapt their schedules in order to reduce the electricity demand in the network in critical times [27]. It would be possible to create a charging station at home and at the workplace or suitable parking spaces where people have to stay longer. Such places are suitable for regular charging while high-speed freeways and shopping malls where the stops are relatively shorter prove to be best for fast charging. It is important that government and private actors work with existing oil companies in setting up stations at present outlets [15].

22.5   DISCUSSING THE PRESENT SCENARIO

For the four-wheel segment, Ola and Uber are going to play a major role in fleet management. It helped in easing traffic congestion by sharing rides and other innovative services. In several studies, it was found that EVs will contribute to a mere 2% of total vehicle sales by 2023. This puts the government's target of 100% penetration by 2030 in a tight spot [2]. Most 3-wheeler auto-rickshaw drivers are not convinced to change because currently any electric 3-wheeler with an almost equal performance of IC motor costs at least twice as much. In addition, driving range anxiety and improper and irregular charging facilities would result in non-operation and loss of income. There is a need to develop a battery replacement infrastructure for cars and buses and a formulated battery exchange standard to ensure safety and functional requirements. In addition, we need to connect public chargers to a computer network for interoperability and proper use [21]. A major concern at this point could be if the nation is ready to meet the goal of complete electrification and stand as a total EV nation while there are areas which still face electricity shortages. The coal is major source of power generation in our country India, which causes air pollution.

22.6   CONCLUSION AND RESULTS

According to the review, the most important factor for the success of EVs is the technological factor. Driving range, power and battery capacity are the most important criteria within this domain. The technological factor is seconded by the environment and surroundings factor which focuses on emissions and associated warming issues. Next comes the economic factor which includes energy and vehicle costs and is of very great practical importance in India. Geographic and cultural factors are comparatively lesser relevant than other factors and do not require immediate attention. The immediate next step is to establish a proper sequence of processes in the fields of energy and transportation to ensure the stability of operations and successful penetration of EVs in India.

REFERENCES

1. B. J. Jiby, R. Shirase. Present and future trends for electric vehicles in India. *CASS*, 3(1) (2019). ISSN: 2581-6403. https://www.scribd.com/document/527943432/presentandfuturetrendsforelectricvehiclesinindia
2. J. P. Kesari, Y. Sharma, C. Goel. Opportunities and scope for electric vehicles in India. *SSRG International Journal of Mechanical Engineering*, 6(5) (2019), 1–8.
3. P. Gode, G. Bieker, A. Bandivadekar. ICCT working paper 2021-07. Battery capacity needed to power electric vehicles in India from 2020 to 2035.
4. J. Brady, M. O'Mahony. Modelling charging profiles of electric vehicles based on real-world electric vehicle charging data. *Sustainable Cities and Society*, 26 (2016), 203–216.
5. P. Morrissey, P. Weldon, M. O. Mahony. Future standard and fast charging infrastructure planning: An analysis of electric vehicle charging behaviour. *Energy Policy*, 89 (2016), 257–270.
6. A. Foley, B. Tyther, P. Calnan, B. O. Gallachoir. Impacts of electric vehicle charging under electricity market operations. *Applied Energy*, 101 (2013), 93–102.
7. S. Steinhilber, P. Wells, S. Thankappan. Socio-technical inertia: Understanding the barriers to electric vehicles. *Energy Policy*, 60 (2013), 531–539.
8. H. Yu, F. Tseng, R. McGee. Driving pattern identification for EV range estimation. In *2012 IEEE International Electric Vehicle Conference*, IEEE, 2012, March, pp. 1–7.
9. J. G. Hayes, R. P. R. De Oliveira, S. Vaughan, M. G. Egan. Simplified electric vehicle power train models and range estimation. In *2011 IEEE vehicle power and propulsion conference*, IEEE, 2011, September, pp. 1–5.
10. F. Salah, J. P. Ilg, C. M. Flath, H. Basse, C. Van Dinther. Impact of electric vehicles on distribution substations: A Swiss case study. *Applied Energy*, 137 (2015), 88–96.

11. N. Hartmann, E. D. Ozdemir. Impact of different utilization scenarios of electric vehicles on the German grid in 2030. *Journal of Power Sources*, 196(4) (2011), 2311–2318.

12. D. Yin, Y. Hori. A new approach to traction control of EV based on maximum effective torque estimation. In *2008 34th Annual Conference of IEEE Industrial Electronics*, IEEE, 2008, November, pp. 2764–2769.

13. L. Lu, X. Han, J. Li, J. Hua, M. Ouyang. A review on the key issues for lithium-ion battery management in electric vehicles. *Journal of Power Sources*, 226 (2013), 272–288.

14. A. Panday, H. O. Bansal. A review of optimal energy management strategies for hybrid electric vehicle. *International Journal of Vehicular Technology*, 2014 (2014), 1–19.

15. A. Kumar, S. K. Choudhary, K. N. Chethan. Commercial viability of electric vehicles in India. *International Journal of Mechanical Engineering and Technology (IJMET)*, 9(6) (2018), 730–745.

16. Niti Aayog - Zero Emission Vehicles (ZEVs): Towards a policy framework, Global Mobility Summit 2018.

17. A. Jhunjhunwala, P. Kaur, S. Mutagekar. Electric vehicles in India. *IEEE Electrification Magazine*, December 2018.

18. P. R. Shukla, S. Dhar, M. Pathak, K. Bhaskar. Electric vehicles scenarios and a roadmap for India. UNEP DTU Partnership, 2014.

19. Future of mobility in India, Envisioning the future of the Indian mobility ecosystem, February 2020.

20. PwC. Indian automotive sector: Creating future-ready organisations, May 2019.

21. White paper on electric vehicles adopting pure electric vehicles: Key policy enablers, December 2017, SIAM.

22. P. Bansal, R. R. Kumar, A. Raj, S. Dubey, D. J. Graham. Willingness to pay and attitudinal preferences of Indian consumers for electric vehicles. *Energy Economics*, 100 (2021), 105340.

23. A. K. Digalwar, R. G. Thomas, A. Rastogi. Evaluation of factors for sustainable manufacturing of electric vehicles in India. *Procedia CIRP*, 98 (2021), 505–510.

24. S. Goel, R. Sharma, A. K. Rathore. A review on barrier and challenges of electric vehicle in India and vehicle to grid optimization. *Transportation Engineering*, 4 (2021), 100057.

25. Harper, G., Sommerville, R., Kendrick, E., Driscoll, L., Slater, P., Stolkin, R., ... & Anderson, P. Recycling lithium-ion batteries from electric vehicles. *Nature*, 575(7781) (2019), 75–86.

26. K. L. Mokariya, V. A. Shah, M. M. Lokhande. Impact of penetration of electric vehicles on Indian power grid. *World Electric Vehicle Journal*, 7 (2015), 518–529. ISSN 2032-6653.

27. P. Kumar, K. Dash. Potential need for electric vehicles, charging station infrastructure and its challenges for the Indian market. *Advance in Electronic and Electric Engineering*, 3(4) (2013), 471–476. ISSN 2231-1297.

Chapter 23

Development of heuristic DSS for supply chain architecture

*Jerin Joseph, Rajeshwar Siddeshwar Kadadevaramath,
Mohan Sangli, B. Lathashankar, Akarsha Kadadevaramath,
and Immanuel Edinbarough*

CONTENTS

23.1 INTRODUCTION

In general, when decisions must be made in a limited and restricted resource setting, optimization problems seek a solution. When one or more of the factors such as Supplier capacity, Production capacity, Warehouse capacity are constrained, supply chain process minimization and maximization problems require matching market requirements and supply of goods. The time required to purchase, create, or transport something is, by far, the most essential restricted resource. Demand cannot be met

DOI: 10.1201/9781003293576-23

instantly due to the limited rate of purchase, manufacturing, distribution, and logistics resources. The demand must always be met with some length of time, and this may not be fast enough unless the supply is generated well ahead of demand. Other resources, such as warehouse stock, space, or vehicle availability, may be constrained in meeting market demand in addition to time. The decision variables and constraints in an optimization problem are mostly as follows.

1. The planner has authority over the following supply chain network (SCN) decision variables:
   • When should you order raw materials from a supplier and how much should you order?
   • When should an order be manufactured?
   • How much and when should the merchandise be shipped to a client or distribution center?
2. SCN constraints are the restrictions that are imposed on the supply plan:
   • The ability of a supplier to generate a list of raw materials or semi-finished components
   • A production line that can run for how many shifts each day, and a worker who can only work a certain amount of overtime
   • The ability of a client or distribution center to record and process the goods receipts

In a SCN, constraints are either rigid or flexible limits that must be adhered to or satisfied – the number of work busy hours in a shift or the maximum limit of a truck. It is possible to relax or violate soft constraints. Customers' due dates or warehouse space limitations are examples of soft constraints. Customer delivery dates can be modified, or a product can be temporarily crammed into a stockyard, easing the limitations.

23.2   GENERAL OPTIMIZATION OBJECTIVES OF THE SUPPLY CHAIN NETWORK

The following are the general optimization objectives of the SCN:

•   Maximizing the company profits or margins
•   Minimizing the total supply chain operating costs
•   Maximizing customer satisfaction and service level
   • Increasing earnings or margins
   • Cutting costs or cycle times in the supply chain
   • Improving customer service
   • Reducing lateness and increasing production throughput

The distinction between a constraint and a goal is sometimes misunderstood by those unfamiliar with optimization. Some elements can be expressed as a goal or a constraint, which adds to the confusion.

23.3   OPTIMIZATION MODELS

These replicas and models describe the connections between choices, restrictions, and goals. These are frequently stated as mathematical languages and variables. In general, the model must accurately replicate the "actual scenario" to capture the requirements of the problem in order to deliver an effective solution and show the main characteristics of the supply chain. A solver/algorithm identifies the optimum plan of action after an optimization issue is formulated. It is computer software that searches for a solution that fulfills the goal using a sequence of logical steps or algorithms.

The following three types of answers can be analyzed and given by a solver:

- **Feasible answer:** It is an answer that satisfies all of the problem's restrictions.
- **Required solution:** It is the best required solution that meets the optimization problem's aim. Though there may be multiple answers and solutions to some scenarios, there is usually only one optimum solution.
- **Very good solution:** It is an answer that only partially meets the optimization problem's goal. It is not the final best answer, but it is satisfactory or reasonable. It is usually one of the best options answer available in the analyzer.

23.4   STATEMENT OF AN OPTIMIZATION PROBLEM

The act of finding the best results for a given task under specific circumstances is known as optimization. Engineers make several decisions during the product design, product manufacturing, and maintenance of the engineering system. The final goal of all such judgments is to reduce the amount of effort required or else increase the desired reward. Finding the limitations that give the maximum or minimum value of a function is the effort needed or the advantage interested in any practical scenario. The following is an example of an optimization problem:

$$\text{Min } f(X) = (X_1, X_2, ..., X_n) \tag{23.1}$$

subject to constraints

$$g_j(X) \leq 0, \quad j = 1, 2, ..., m;$$

$$I_j(X) = 0, \quad j = 1, 2, ..., p;$$

The problem stated in Eq. (23.1) is called a constrained optimization problem. Some optimization problems do not involve any constraints and are called unconstrained optimization problems (see Eq. 23.2). An unconstrained optimization problem can be stated the same as 23.1 but without constrains.:

$$\text{Min } f(X) = (X_1, X_2, ..., X_n) \tag{23.2}$$

Most of the practical problems are constrained in nature. The following are some of the reasons why studying unconstrained problems is important:

a.   Constraints have no substantial impact on certain design and manufacturing challenges.
b.   Unconstrained minimization approaches are required by some of the most powerful and reliable strategies for handling constrained optimization issues.
c.   Comprehending unconstrained minimizing techniques provides the foundation for understanding constrained minimization methods.
d.   Most of the constrained optimization problems can be converted into unconstrained problems using methods like the penalty function approach and are solved using unconstrained algorithms.

23.5 LITERATURE REVIEW ON MULTI-STAGE SUPPLY CHAIN ARCHITECTURE MODELS

The cross-functional components of the supply chain are highlighted in a supply chain management literature. It is a significant task to identify the perfect plan that works along the whole supply chain [1,2]. A supply chain viewpoint gives the prospect for large inventory savings through efficient coordination and right scheduling of purchase, manufacture, and distribution of items along the SCN, according to an emerging supply chain management paradigm. Supply chains [3] are "real-world scenarios that convert raw materials and semi-furnished products into end goods that are marketed to customers". Supply chains consist of many phases that create value by transforming and consuming time, place, and material.

On a finite horizon [4], a certainty model for calculating minimum inventory levels and procurement times associated with the lowest cost solutions was established for a collaborated supply chain. Based on economic order quantity methodologies [5], provide a Mixed Integer Programming (MIP) Models and nonlinear mathematical programming models and are analyzed a restricted optimization model called PILOT to examine the outcome of various factors on supply chain cost, as well as the other difficulty of deciding which production and distribution centers should be open. Green Supply Chain Management (GSCM) Model [6], a MIP model for evaluating continental supply chain options consisting of multiple stages (echelons). To minimize working time and expenses related to manufacturing, stock, and transportation, GSCM has considered the interlinking of production, inventory, and delivery operations. The proposed model and solution techniques are easy to apply but restricted to a single time period and no limitations. Supply chain architectures with many stages consist of flow of goods among raw material suppliers, goods producers, break-bulk points or centers and customers. A near-optimal inventory cost function was established and was incorporated into a fixed facility location model [7]. This one model was developed to take a trade-off between market requirement coverage and the cost involved in locating vehicle warehouses.

23.5.1   Motivation for the study

Most of the supply chain models considered demand as certain and also considered only two tiers or focused mainly on the supply chain's middle operational level. Based on the study, it is found that there are still many factors that affect the supply chain profitability and are missing in the existing models on the design of SCNs.

As a result, the goal of Supply Chain Management (SCM) is to determine the best or near-best alternative architecture for meeting high-level efficiency in the supply chain. Supply chain operations must run in cross-functional and mutual cooperation to achieve high-level performance. There are a number of difficult issues that come with designing a collaborated and integrated supply chain.

23.6   PROBLEM STATEMENT AND RESEARCH OBJECTIVES

The ability of firms to compete and differentiate themselves is hampered by weekly collaborated enterprise logistic components and procedures in SCM. Management is interested in finding and eliminating the causes of inefficiency and waste by using an integrated approach to supply network architecture. As a result, a computer simulation study can aid in the investigation and evaluation of network setup problems. Traditional optimization algorithms are used to solve smaller-sized problems. Actually, the real-world scenarios and issues are highly complicated and combinatorial in nature. To solve such kinds of problems, generally, heuristic approaches are used.

23.6.1   Research objectives

Traditional analysis has assumed that the enterprise is loosely linked, that is, distinct subsystems operate largely independently of one another. The topic of integrated modeling and analysis is addressed using a process-oriented approach. It is becoming more applicable as material, information, and service flows in today's manufacturing organizations become more tightly integrated. Some frameworks and models for designing and analyzing supply chain operations have been established [8]. These existing models, on the other hand, are either oversimplified or only qualitatively defined (some are only based on simulation studies), making them difficult to apply to real-world supply chains for quantitative analysis and decision-making. Furthermore, there have been few studies that have focused on integrated modeling and analysis that may be utilized to assess supply chain throughput capacity across different stages. Because of the enormous increase in modeling complexity required, previous research have overlooked the significant consequences of these integration difficulties. As a result, the competence and applicability of models developed in previous studies to assess real supply chain business operations are severely constrained.

The following study objectives are based on the findings of the literature review and are aimed at delivering and creating efficient outcomes for performance analysis of collaborated supply chain architecture. The following goals are pursued in this research work:

- To develop a mathematical (integer programming) model for three echelon SCN.
- To develop and Non-Linear Inertia Weight Particle Swarm Optimization (NLIW-PSO) Algorithm
- To conduct performance analysis for three echelon SCN architecture and obtain the optimal solution for total SCN profit, which is used as a supply chain performance indicator.

23.7 OPTIMIZATION OF THREE-STAGE SUPPLY CHAIN ARCHITECTURE USING NLIW-PSO ALGORITHM

This work discusses the swarm particle representation, velocity determination of all particles, working of Particle Swarm Optimization (PSO) algorithm, experimental design and results and discussions of supply chain architecture optimization.

23.7.1 Particle development in PSO algorithm of three-stage supply chain architecture

In the PSO algorithm, one solution in supply chain configuration is shown by a particle. It is nothing but one string of integers called decision variables. Depending on the swarm number or size (say k), there will be N particles signifying the SCN configuration.

23.7.2 Velocity determination and position modification equations used for the optimization of SCN architecture

The NLIW-PSO is used in this research study. The following are the equations of NLIW-PSO used for velocity determination and position modification of particles of the PSO.

$$\text{Velocity}, (v)_{kd}^{\text{new}} = w_{\text{iter}} \times v_{kd} + c_1 \times \left[r_1 \times \left(P_{kd} - X_{kd} \right) c_2 \right] \times \left[r_2 \times \left(G_d - X_{kd} \right) \right] \qquad (23.3)$$

$$w_{\text{iter}} = \left\{ \frac{\left(\text{iter}_{\max} - \text{iter} \right)^n}{\left(\text{iter}_{\max} \right)^n} \right\} w_{\text{initial}} - w_{\text{final}} + w_{\text{final}} \qquad (23.4)$$

$$m = \frac{\left(w_{\text{initial}} - w_{\text{final}} \right)}{\text{iter}_{\max}} \qquad (23.5)$$

$$w_{\text{final}} = w_{\text{initial}} + m \times \text{iter}_{\max} \qquad (23.6)$$

$$X_{kd}^{\text{new}} = X_k + v_{kd} \qquad (23.7)$$

23.7.3   Structure of NLIW-PSO algorithm

Here, 'k' denotes the number of particles (Figure 23.1).

'D' denotes the maximum number of dimensions within the minimum and maximum limits for each dimension.

$\{P_{kd}\}$(best point of the particle, i.e., particle best).

$\{G_d\}$(globalbest).

Step1:   Initializing the particle position $\{X_{kd}, d=1,2,\ldots,D\}$

$$\text{set}^W\text{iter} = \left\{ \frac{|(\text{iter}_{max} - \text{iter})^n|}{(\text{iter}_{max})^n} \right\} (W_{initial} - W_{final}) + W_{final}$$

Step2:   Initialize the particle velocity $\{v_{kd}, d=1,2,\ldots,D\}$

Step3:   Calculate the maximum velocity of the particles
          $v_{max}=0.5 \times$ Maximum limit for each dimension

Step4:   if $v_{kd} > v_{max}$
          set $v_{kd} = v_{max}$ for all 'k' and 'd'

Step5:   find $Z\{X_{kd}\}$
          Initialize $\{P_{kd}\}$ and $\{G_d\}$
          Go to Step 7.

Step6:   Update $\{P_{kd}$ and $\{G_d\}$
          If $\text{iter}_{max}=2000$ or penalty=0;
          Goto Step 10

Step7:   Calculate new velocity $v^{new}{}_{kd}$

$$(v_{kd})^{new} = W_{iter} \times v_{kd} + c_1 \times \left[r_1 \times (P_{kd} - X_{kd}) \right] + c_2 \times \left[r_2 \times (G_d - X_{kd}) \right]$$

          Check if
          $v_{kd}^{new} > v_{max}$
          set $v_{kd}^{new} = v_{max}$ for all 'k'and'd'.

Step8:   new position of the particle
          $X_{kd}^{new} = v_{kd}^{new} + X_{kd}$
          $SetX_{kd}^{new} = X_{kd}$
          Find $Z\{X_{kd}\}$

Step9:   iter=iter+1
          Go to Step 6

Step10:   Print $\{G_d\}$ and $Z\{G_d\}$;Stop

Figure 23.1 Structure of NLIW-PSO algorithm for SCS.

23.7.4   Development of initial set particles in the PSO algorithms

Generate K (Swarm Size) solutions $\{X_{kd}, d = 1,2,...,D\}, (k = 1,2,...,5)$ as follows:

Set $X_{kd} = LL_d + (UL_d - LL_d)*U(0,1)$ for $d = 1,2,...,D$

23.7.5   Generation of initial velocities in the proposed PSO algorithms

Generate K (Swarm Size) solutions $\{v_{kd}, d = 1,2,...,D\}, (k = 1,2,...,5)$ as follows:

Set $v_{kd} = LL_d + (UL_d - LL_d)*U(0,1)$ for $d = 1,2,...,D$.

where with respect to dimension space d,

LL_d = Minimum limits on the value of X_k
UL_d = Maximum limits on the value of X_k
$U (0, 1)$ = Uniformly distributed random number in the range (0, 1).

23.8   PERFORMANCE ANALYSIS OF SCN ARCHITECTURE

23.8.1   Experimental design

The supply chain configurations are studied in this computational study to determine the performance. The Supply Chain Setting (SCS) was optimized with the goal of increasing the SCN profit of the company using PSO algorithms. Different raw material purchase costs, constraints, logistics costs, inventory holding costs, and selling prices are included in these supply chain test setups to evaluate proposed algorithms' robustness with respect to consistency and better performance across SCNs. The performance of NLIW-PSO is evaluated by taking into account the customer market requirement as uniformly distributed in the range [100, 200] and a random number generator is applied to produce the uniformly distributed random numbers. Fifteen iterations are considered to have the results. The PSO algorithm is written in C language.

23.8.2   Pilot studies for parameters settings

To determine the best values of parameters of NLIW-PSO, pilot studies have been conducted in this research.

23.8.3   Results and discussions

The performances of the SCN and the PSO algorithms were evaluated by considering the supply chain test problems. Twenty market requirement scenarios are taken. The performance of the proposed NLIW-PSO algorithm was tested. Near-optimal solutions were obtained, after running the PSO algorithm to a maximum number of iterations, as reported in Tables 23.12 to 23.19. Table 23.12 also gives the mean, best and worst values of total supply chain cost (TSCC) and profit over 15 replications for each scenario. The decision variables corresponding to the best optimal solutions obtained by NLIW-PSO are presented in Table 23.13(a). Another important key determinant of measure of effectiveness to evaluate the performance of the solution procedure is the average computational effort required by an algorithm. The time (in seconds) that an algorithm has enumerated to converge to the near-optimal solutions is given in Table 23.14.

23.8.4 Supply Chain Setting (SCS)

Vendors (V) = 1…3 in numbers
    Components (C) = 1…3 in units
    Plants (P) = 1…2 in numbers
    Warehouses (W) = 1…3 in numbers
    Distribution Centers (DC) = 1…6 in numbers

Please refer to Tables 23.1 to 23.12.

Table 23.1 Capacity of vendors

	Component 1 (units)	Component 2 (units)	Component 3 (units)
V1	600	500	900
V2	400	400	300
V3	600	700	400

Note: V1=Vendor1; V2=Vendor2; V3=Vendor3

Table 23.2 Cost of the raw material component at the vendor (in Rs/unit)

	C1	C2	C3
V1	600	415	250
V2	530	380	220
V3	570	395	235

Note: V1=Vendor1; V2=Vendor2; V3=Vendor3
C1=Component1; C2=Component2; C3=Component3

Table 23.3 Transportation cost of Component 1 (in Rs/unit)

	P1	P2
V1	20	24
V2	30	35
V3	25	28

Note: V1=Vendor1; V2=Vendor2; V3=Vendor3
P1= Plant1; P2=Plant2

Table 23.4 Transportation cost of Component 2 (in Rs/unit)

	P1	P2
V1	15	12
V2	16	17
V3	13	15

Note: V1=Vendor1; V2=Vendor2; V3=Vendor3
P1= Plant1; P2=Plant2

Table 23.5 Transportation cost of Component 3 (in Rs/unit)

	P1	P2
V1	8	10
V2	11	12
V3	13	10

Note: V1=Vendor1; V2=Vendor2; V3=Vendor3
P1= Plant1; P2=Plant2

Table 23.6 Plant transportation cost (in Rs/product)

	W1	W2	W3
P1	100	85	90
P2	110	105	95

Note: P1= Plant1; P2=Plant2
W1=Warehouse1; W2=Warehouse2; W3=Warehouse3

Table 23.7 Warehouse transportation cost (in Rs/product)

	DC1	DC2	DC3	DC4	DC5	DC6
W1	18	22	26	30	32	35
W2	44	35	30	33	38	40
W3	47	45	40	38	32	30

Note: W1=Warehouse1; W2=Warehouse2; W3=Warehouse3
DC1=Distribution Center1; DC2=Distribution Center2;
DC3=Distribution Center3; DC4=Distribution Center4;
DC5=Distribution Center5; DC6=Distribution Center6

Table 23.8 Data related to plants

	PI	P2
Capacity of plant 'p'	600	900
Manufacturing cost of plant (in Rs/product)	3,000	3,400
Inventory cost of plant (in Rs/product)	100	125

Table 23.9 Data related to warehouses

	WI	W2	W3
Capacity of warehouse (units)	400	600	400
Inventory cost of distribution center (in Rs/product)	110	100	95

Table 23.10 The selling price at each distribution center

	DCI	DC2	DC3	DC4	DC5	DC6
Selling price (Rs)	7,000	7,500	8,000	7,300	8,200	7,700

Table 23.11 Demand for SCS

Scenario	Demand U (100, 200) at different distribution centers					
	DCI	DC2	DC3	DC4	DC5	DC6
1	166	186	168	105	114	186
2	142	177	142	140	133	105
3	184	119	118	172	173	121
4	188	146	158	149	159	113
5	200	141	154	162	171	172
6	159	160	178	144	162	161
7	194	198	101	103	109	118
8	136	159	151	130	181	133
9	127	196	138	166	172	156
10	177	150	142	147	169	133
11	168	199	195	109	154	176
12	174	132	199	138	178	113
13	188	171	108	175	127	104
14	187	117	101	198	187	166
15	143	163	148	104	102	113
16	172	197	146	156	115	110
17	125	145	174	123	196	175
18	165	194	189	116	112	153
19	134	122	163	146	172	149
20	183	185	177	188	163	156

Table 23.12 Results of performance evaluation-profit as a performance indicator by NLIW-PSO

Scenario	1	2	3	4	5	6	7	8	9	10	11	12	13	14	15	Best	Worst	Mean	STD
1	1473453	2632538	1882731	2292285	2480653	2389557	2284102	1851918	1871069	2013377	2808025	1777982	2395506	1850656	1860703	2808025	1473453	2124304	372954
2	2423249	1504870	362862	1882280	1132080	1656589	816134	915108	949256	838970	1799282	1365636	1780335	2058022	1312487	2423249	362862	1386477	556842
3	1762763	1680110	2219688	1986834	1319859	1543338	1589271	1852805	2039572	1563507	1973203	1924208	2333963	1206278	2212968	2333963	1206278	1813891	331062
4	1353889	1952869	1438853	1875723	1873195	1740334	1670933	1997258	2137059	2122504	2009332	2149821	1793616	2638369	1636153	2638369	1353889	1892662	317108
5	1143152	2359187	2496418	2321792	2304847	2501793	2230039	2397774	1811990	2786951	2395058	2572415	2653117	2658442	2502452	2786951	1143152	2342362	401943
6	1258667	2156772	2542792	1591898	2132283	2278831	2216927	1577764	2369379	2233256	2417943	1931713	2885009	2560413	2126581	2885009	1258667	2152015	421968
7	1344527	924756	994934	1831156	1220864	2023574	1615648	1069993	583947	1362043	2294911	499301	1766757	1658950	1110377	2294911	499301	1353449	513232
8	1465120	2292802	1786348	1950856	1474250	2234949	2170071	1459002	2213058	1754418	2439082	2494242	2659183	1891869	1206813	2659183	1206813	1966138	437039
9	2015019	2443682	1594522	2496679	1978542	2540959	2132543	2268425	2297169	2152129	2571519	2638340	2430118	2122800	2237766	2638340	1594522	2261347	276268
10	1674225	2258946	2132859	2121194	2399730	1997385	1710196	2060132	1347145	2438321	2351015	2468320	2193466	1784728	1447329	2468320	1347145	2025666	357884
11	2425938	2494778	2277834	2576708	2455758	2376433	2960879	2468511	2683588	2152269	2283039	2519007	2662187	2565903	2553833	2960879	2152269	2497111	193991
12	1036501	1965776	2240001	1766698	2031344	2134420	2187259	2392306	2073201	2218284	1961168	2117960	2636793	1881476	2460160	2636793	1036501	2073556	365291
13	1119828	1760787	841056	1719846	1787138	1962977	1162190	1510275	1350173	833919	1894904	2233637	1979648	2462379	1859870	2462379	833919	1631908	484851
14	1155580	1893804	1532878	2426234	1683419	2472730	2089209	2409332	1661127	1673914	2807924	2155583	2261449	2066998	1652112	2807924	1155580	1996153	440387
15	946539	1123694	1389904	907515	216639	1731757	976561	1157334	1076888	1447534	1836808	1229283	1446107	1816602	928251	1836808	216639	1215428	421481
16	1651670	1826035	1749245	1579107	1970122	2007136	1526538	1954706	1509121	1162547	2258043	2039329	2316319	2206706	860781	2316319	860781	1774494	405940
17	906755	2188727	1534192	2699107	1724718	2085384	1729681	2102066	2270085	2050340	2445564	1958873	2189971	1818460	1677691	2699107	906755	1958828	424830
18	1921810	2347687	2378987	2166591	2452566	2148031	2662007	2186945	1753742	998914	2428700	2653948	1852437	2016798	2321025	2662007	998914	2152679	418160
19	770209	2426325	1684360	2276707	1471654	1497591	2243861	1609589	1806442	1963529	2343130	2319636	2304351	2061114	1574844	2426325	770209	1890223	460500
20	1728060	2647835	2650987	2881201	2609922	2922332	2862984	2631912	2653939	2684075	2981445	2843191	2737955	2746125	2682797	2981445	1728060	2684317	289536

Table 23.13a Decision variables for best near-optimal solutions

Scenario	x111	x112	x121	x122	x131	x132	x211	x212	x221	x222	x231	x232	x311	x312	x321	x322	x331	x332	Y11	Y12	Y13	Y21	Y22	Y23
1	65	221	186	93	257	103	60	177	36	2	412	238	282	339	225	66	6	13	205	117	186	83	180	154
2	385	0	1	148	3	362	252	54	0	219	79	237	368	0	2	229	9	282	68	163	100	243	161	105
3	**305**	74	57	330	275	98	96	366	48	64	401	98	396	269	102	88	190	152	307	53	172	15	260	80
4	108	281	63	220	325	0	109	266	1	228	432	0	125	125	1	166	269	126	378	27	91	22	201	194
5	353	205	3	164	53	222	0	378	246	120	201	346	472	250	169	243	0	98	180	139	90	119	166	306
6	224	128	81	295	318	0	2	167	17	106	577	95	264	453	33	0	177	0	125	200	271	141	189	38
7	289	239	128	79	66	45	159	339	160	68	167	145	394	161	99	83	56	129	227	209	47	100	132	110
8	227	144	34	333	152	2	0	359	0	115	517	6	292	229	33	198	21	51	132	115	165	163	153	162
9	142	172	370	30	33	409	243	246	44	67	280	119	238	523	22	102	324	63	239	171	135	79	143	188
10	398	82	126	56	108	382	354	57	67	104	332	182	261	124	240	238	317	56	355	132	107	58	148	134
11	201	178	142	228	145	187	229	230	136	118	127	185	139	339	35	49	121	146	105	66	317	95	368	50
12	285	120	54	229	248	82	69	167	81	233	568	60	358	229	34	265	110	60	163	112	228	192	111	128
13	344	184	0	303	164	3	7	261	10	2	384	214	57	234	2	245	310	4	346	39	11	54	146	277
14	161	195	168	69	275	92	1	361	300	0	300	0	668	3	139	160	199	200	88	183	329	59	234	63
15	102	249	227	103	109	116	24	430	134	159	224	147	251	308	42	87	188	198	149	211	14	81	136	201
16	164	92	41	109	380	207	131	69	67	316	580	93	527	254	109	192	88	77	105	267	163	104	121	145
17	194	340	152	0	132	146	383	35	61	92	188	361	552	0	139	144	1	342	84	200	169	223	105	158
18	187	92	95	274	162	132	148	135	93	29	229	437	184	303	201	130	128	143	17	128	290	155	229	114
19	142	107	34	280	275	123	58	192	204	111	225	215	152	487	1	41	59	283	216	169	26	146	229	112
20	177	371	203	68	146	133	74	369	299	99	145	210	490	144	1	240	43	223	161	246	81	186	64	322

Table 23.13b Decision variables for best near-optimal solutions of SCS-II

Scenario	Z11	Z12	Z13	Z14	Z15	Z16	Z21	Z22	Z23	Z24	Z25	Z26	Z31	Z32	Z33	Z34	Z35	Z36
1	114	27	67	61	12	7	34	82	79	14	16	72	18	77	22	30	86	107
2	22	27	32	80	124	26	98	112	72	38	2	1	22	38	38	22	7	78
3	58	33	29	87	63	52	10	39	77	59	93	35	116	47	12	26	17	34
4	98	61	17	100	45	79	11	45	63	38	51	20	79	40	78	11	63	14
5	144	51	45	47	0	12	35	34	43	36	7	150	21	56	66	79	164	10
6	70	4	30	93	44	25	61	95	40	40	113	40	28	61	108	11	5	96
7	118	24	67	32	67	19	49	84	24	54	39	91	27	90	10	17	3	8
8	113	18	84	13	22	45	18	119	27	62	18	24	5	22	40	55	141	64
9	39	64	46	109	2	58	38	50	66	53	49	58	50	82	26	4	121	40
10	9	39	94	119	25	114	146	58	34	25	0	14	22	53	14	3	144	5
11	66	5	–	78	41	9	102	53	190	10	59	20	0	141	4	21	54	14
12	28	20	105	83	32	87	52	65	11	47	46	2	94	47	83	8	100	24
13	83	101	0	170	0	46	56	38	22	5	26	38	49	32	86	0	101	20
14	27	9	13	27	2	69	109	–	35	141	58	73	51	107	53	30	127	24
15	57	33	10	30	98	0	76	41	102	–	2	108	10	89	36	73	2	5
16	27	69	49	4	59	0	42	98	68	133	36	3	103	30	29	19	20	107
17	81	90	62	29	2	43	28	23	83	22	47	–	16	32	29	72	147	31
18	51	4	27	67	19	4	33	151	130	24	13	101	81	39	32	25	80	143
19	48	56	87	0	103	68	20	52	50	145	50	6	66	14	26	–	19	8
20	135	6	77	26	71	27	45	24	21	136	15	69	3	155	79	26	77	60

Table 23.14 Computational effort in obtaining near-optimal solution

Scenario	Computational effort in seconds for each optimal solution															
	1	2	3	4	5	6	7	8	9	10	11	12	13	14	15	Average
1	2	1	1	2	3	4	3	2	1	3	1	4	1	1	2	2
2	1	1	4	2	1	5	34	2	2	2	3	4	1	2	2	4
3	1	2	1	2	3	4	1	25	1	1	3	4	1	4	1	4
4	2	1	3	2	2	2	3	1	4	1	4	4	1	2	5	2
5	2	1	1	2	2	2	2	2	2	2	1	1	3	2	3	2
6	2	1	2	2	1	1	2	3	3	5	3	7	2	2	7	3
7	1	1	2	2	2	8	1	2	2	1	2	4	1	4	1	2
8	1	2	1	2	1	5	1	5	1	1	2	2	2	1	3	2
9	1	2	1	1	2	2	1	2	2	1	1	3	1	1	2	2
10	2	2	1	8		1	4	1	2	3	1	4	2	3	4	3
11	1	2	2	1	8	1	4	1	2	3	1	4	2	2	3	2
12	1	1	1	2	1	2	3	3	1	1	4	3	2	3	3	2
13	1	2	1	2	3	2	4	2	2	2	2	2	2	3	3	2
14	1	3	2	2	5	4	3	2	1	4	5	1	1	3	2	3
15	2	1	2	2	3	2	1	2	2	1	2	4	1	3	4	2
16	1	1	2	3	3	1	1	4	1	1	2	2	1	2	2	2
17	2	1	3	2	3	4	7	6	2	3	3	3	1	2	4	3
18	2	3	3	2	2	2	1	2	2	1	1	7	1	1	6	2
19	2	1	2	2	5	4	1	1	3	1	1	4	2	1	2	2
20	1	2	2	3	2	3	1	2	3	2	3	3	1	5	2	2

For SCS:

Table 23.15 Various optimal supply chain cost components

Total sales (Rs)	TSCC (Rs)	Profit (Rs)	TSC (Rs)	TMC (Rs)	TDC (Rs)	TTC (Rs)
7,034,500	4,226,475	2,808,025	1,117,950	2,942,425	29,826	136,274

Note: TSC=Total Supply Cost; TMC=Total Manufacturing Cost; TDC=Total Distribution Cost; TTC= Total Transportation Cost

Table 23.16 The optimal distribution of the components (in units)

	Component 1		Component 2		Component 3	
	Plant 1	Plant 2	Plant 1	Plant 2	Plant 1	Plant 2
Vendor 1	65	221	60	177	282	339
Vendor 2	186	93	36	2	225	66
Vendor 3	257	103	412	238	6	13

Table 23.17 The optimal product distribution to warehouses

	Warehouse 1 (units)	Warehouse 2 (units)	Warehouse 3 (units)
Plant 1	205	117	186
Plant 2	83	180	154

Table 23.18 The optimal product distribution to warehouses

	DC1 (units)	DC2 (units)	DC3 (units)	DC4 (units)	DC5 (units)	DC6 (units)
Warehouse 1	114	27	67	61	12	7
Warehouse 2	34	82	79	14	16	72
Warehouse 3	18	77	22	30	86	107

23.9 CONCLUSION

The present work considered the mathematical modeling of three echelon supply chain network and application of Particle Swarm Optimization algorithm for the best alignment of procurement, production and distribution in three echelon supply chain network in order to optimize profit in supply chain network optimization.The algorithm called non-linear inertia weight (NLIW-PSO) PSO, where particle velocity is updated dynamically. Extensive performance analysis using the PSO variants has been carried out on real-word SCN problems. Results indicates non-linear inertia weight PSO algorithm (NLIW-PSO) generating better quality solutions to the problem considered in this study. The better performance of the above solution methodology is due to the novel solution construction procedure implemented in the algorithm.

Table 23.19 Supply chain components for best near-optimal solutions

Scenario	Demand U (100, 200)						TSC	TSC % of TSCC	TMC	TMC% of TSCC	TWC	TWC% of TSCC	TTC	TTC % of TSCC	TSCC	Revenue	Profit	TSCC% of Rev	Profit % of Rev
	R1	R2	R3	R4	R5	R6													
1	166	186	168	105	114	186	1117950	26	2942425	70	29826	0.71	136274	3.22	4226475	7034500	2808025	60	40
2	142	177	142	140	133	105	1064255	27	2734700	69	28626	0.72	127770	3.23	3955351	6378600	2423249	62	38
3	184	119	118	172	173	121	1342935	31	2879650	65	30565	0.7	143287	3.26	4396437	6730400	2333963	65	35
4	188	146	158	149	159	113	1196210	28	2930525	68	30181	0.7	141315	3.29	4298231	6936600	2638369	62	38
5	200	141	154	162	171	172	1349285	28	3278225	68	32885	0.68	151354	3.15	4811749	7598700	2786951	63	37
6	159	160	178	144	162	161	1232080	28	3060800	68	33463	0.75	144948	3.24	4475291	7356300	2881009	61	39
7	194	198	101	103	109	118	1114075	28	2644850	68	27774	0.71	123690	3.16	3910389	6205300	2294911	63	37
8	136	159	151	130	181	133	116670	27	2872175	69	28087	0.68	133685	3.22	4150617	6809800	2659183	61	39
9	127	196	138	166	172	156	1365955	29	3098325	67	32307	0.7	150973	3.25	4648060	7286400	2638340	64	36
10	177	150	142	147	169	133	1343440	30	2990925	66	33264	0.74	147051	3.26	4514680	6983000	2468320	65	35
11	168	199	195	109	154	176	1268810	27	3224925	69	34147	0.73	153439	3.28	4681321	7642200	2960879	61	39
12	174	132	199	138	178	113	1299110	29	3023300	67	32582	0.72	145315	3.23	4500307	7137100	2636793	63	37
13	188	171	108	175	127	104	1132490	27	2824375	69	28568	0.69	134388	3.26	4119821	6582200	2462379	63	37
14	187	117	101	198	187	166	1228285	28	3039300	68	35225	0.79	140766	3.17	4443576	7251500	2807924	61	39
15	143	163	148	104	102	113	1240225	31	2638675	65	29139	0.72	128353	3.18	4036392	5873200	1836808	69	31
16	172	197	146	156	115	110	1347570	30	2941875	66	31314	0.7	141222	3.17	4461981	6778300	2316319	66	34
17	125	145	174	123	196	175	1282150	28	3052950	68	30390	0.67	142503	3.16	4507993	7207100	2699107	63	37
18	165	194	189	116	112	153	1202040	27	3026825	69	31651	0.72	142777	3.24	4403293	7065300	2662007	62	38
19	134	122	163	146	172	149	1233290	28	2947750	68	30677	0.7	142458	3.27	4354175	6780500	2426325	64	36
20	183	185	177	188	163	156	1380800	28	3437825	69	36192	0.72	158438	3.16	5013255	7994700	2981445	63	37
Min	125	117	101	103	102	104	1064255	26	2638675	65	27774	0.67	123690	3.15	3910389	5873200	1836808	60	31
Max	200	199	199	198	196	186	1380800	31	3437825	70	36192	0.79	158438	3.29	5013255	7994700	2981445	69	40
Avg	166	163	153	144	152	141	1242881	28.25	2979520	67.8	31343	0.71	141500	3.22	4395470	6981585	2586115	63.05	36.95
STD	23	28	29	28	30	27	96911.7	1.37	196462	1.4	2389.2	0.03	8880.6	0.05	279576	500804	273211.2	2.09	2.09

Note: Total warehouse Cost

Min=Minimum; Max=Maximum; Avg=Average; STD Standard Deviation

REFERENCES

1. Simchi-Levi D., Kaminsky D., Simchi-Levi E. (1999) *'Designing and Managing the Supply Chain: Concepts, Strategies and Case Studies'*. New York: Mcgraw-Hill.
2. Quinn F.J. (2000) 'The Master of Design: An Interview with David Simchi-Levi'. *Supply Chain Management Review*, Vol. 4, pp. 74–80.
3. Hicks D.A. (1999) 'Four-Step Methodology for Using Simulation and Optimization Technologies in Strategic Supply Chain Planning'. *Proceedings of the 1999 Winter Simulation Conference*, Vol. 2, pp. 1215–1220.
4. Ishii K., Takahashi K., Muramatsu R. (1988) 'Integrated Production, Inventory and Distribution Systems'. *International Journal of Production Research*, Vol. 26, No. 3, pp. 473–482.
5. Cohen M.A., Lee H.L. (1989) 'Resource Deployment Analysis of Global Manufacturing and Distribution Networks'. *Journal of Manufacturing and Operations Management*, Vol. 2, pp. 81–104.
6. Arntzen B.C., Brown C.G., Harrison T.P., Trafton L.L. (1995) 'Global Supply Chain Management at Digital Equipment Corporation'. *Interfaces*, Vol. 25, pp. 69–93.
7. Nozick L.K., Turnquist M.A. (2001) 'Inventory, transportation, service quality and the location of distribution centers'. *European Journal of Operational Research*, Vol. 129, pp. 362–371.
8. Vernadat F.B. (1996) *'Enterprise Modeling and Integration: Principles and Applications'* Chapman & Hall, London.

Index